STATISTICAL OBSERVATIONS OF POSITIONING, NAVIGATION, AND TIMING IN A COMBAT SIMULATION

I. Introduction

Imagine a situation where a well-placed smart bomb could end a bloody and expensive conflict. The target is placed near a civilian structure where a miss could cause serious non-combatant damage resulting in moral and political fallout. Spoofing and jamming of Global Positioning System (GPS) signals could relegate the smart bomb to an inelegant piece of shrapnel without regard for location or consequences.

GPS has become as important as any functioning system used today. With many uses, it has become irreplaceable for civilian to military users alike.

1.1 Background of the Study

The GPS is broken down into three components: space, control, and user. The space portion is made up of 32 operational satellites. The US has a consensus effort to maintain at least 24 operational GPS satellites, 95% of the time. These satellites orbit in the medium Earth orbit (MEO) circling Earth twice a day. [1]

The next component is the control component. This component is broken down into the Master Control Station (MCS), the monitor stations, and the ground antennas. The MCS, placed at Colorado Springs, CO, is the central control node for the GPS satellite constellation. It is responsible for all parts of constellation command and control. The monitor stations, located around the world, continuously collect GPS data, which is then routed back to the MCS and merged with other parameters to generate the Navigation

Message. Each satellite is seen from at least three monitor stations, allowing for improved system accuracy. The final portion of the control component is the ground antennas. They uplink satellite orbits (ephemerides) and clock correction information within the Navigation Message and command telemetry from the MCS back to the satellites. The information is updated at least daily. [2]

For this study, attacks against GPS are limited to spoofing and jamming. These attacks target GPS codes, Coarse/Acquisition (C/A) code for civilians, and the restricted Precision (P) code for military usage. It is easier to attack the C/A code (1023 bits long) with its 1.023 Mbit/s repeating every millisecond compared to the P code (6.1871×10^{12} bits long) with its 10.23 Mbit/s repeating once a week. [3]

Spoofing is misleading the receiver with fake signals for positioning calculations, which result in an increase in measured distance. There are three classifications of spoofing attacks. The most basic is the simplistic spoof. This uses a GPS signal simulator and transmitter to overload another GPS receiver with the original and spoofed signal. The intermediate spoof synchronizes a generated GPS signal with current GPS broadcast satellite signals in view [4]. It attacks each channel of the receiver and forces the tracking loops to lock on to the spoofed signals. Sophisticated spoofing, using multiple transmitting antennas, spoofs the current broadcast satellites in addition to spoofing other spoofers' signals. [5]

The easier of the two attacks due to the weak power of GPS signals is jamming interference. Jamming interference is comprised of impulse train, single and multi-tone Continuous Wave (CW), frequency-hop and linear chirp CW. Its main purpose is to prevent the receiving and/or sending of signals. A signal jammer sends a noisy signal on the same frequency to the targeted GPS unit preventing transfer of information.

1.2 Problem Definition

GPS is such an important aspect of the daily lives of not only civilians, but also more for military personnel, attacks against its infrastructure could cause catastrophic damage against not only the user, but others as well. This paper studies the impact of GPS manipulation caused by jamming and spoofing methods with the aid of computer simulation. The objective of the study is to answer the following questions:

- What has a higher effect on a GPS mission, jamming or spoofing?

- Can a predictive model be established for Measures of Effectiveness?

1.3 Scope

Simulation is the tool used for this research. To answer the problem definition, an established model used to observe PNT is utilized. This model uses the System Effectiveness and Analysis Simulation (SEAS) pitting a blue force versus a red force. Data retrieved from this scenario is analyzed using statistical techniques to determine the answers to the problem definition.

1.4 Thesis Overview

The following chapters are the Literature Review, Methodology, Analysis, and Conclusions. The Literature Review discusses related studies about GPS, spoofing, jamming, and simulation. The Methodology chapter discusses the analysis technique that is employed. The simulation model in detail and the statistical approaches that are used for better understanding the model. Chapter 4 presents the data obtained from running the simulation using the statistical method of Design of Experiments method. It looks to answer the questions posed in the problem definition. The final chapter looks to conclude the work by summarizing the methodology and analysis. It also recommends improvements and future research opportunities.

II. Review of Related Studies and Literature

2.1 Introduction

This chapter provides a brief background from open source material on spoofing, jamming, simulation, and statistical design of experiments as they apply to the modeling and analysis of the U.S. NAVSTAR system. We also discuss the combat modeling simulation tool selected for this research along with some related studies and conclude with an introduction to the combat scenario we are studying.

2.2 Global Positioning System (GPS)

NAVSTAR GPS provides positioning, navigation, and timing (PNT) through a 32 linked satellite system flying in six equally spaced medium Earth orbit (MEO) orbital planes circling the Earth twice a day. The system is broken into three components, space, control, and user.

The United States Air Force (USAF), with a commitment that 95% of the time there are at minimum 24 operational GPS satellites, develops, maintains, and operates the space component constellation. The solar powered satellites orbit at approximately 12,550 miles reaching speeds of 7,000 mph. The orbital planes housing the GPS satellites contain four seats occupying baseline satellites. When three satellites are "seen" by a GPS receiver, a 2D position (latitude and longitude) can be determined. With four satellites, a 3D position (2D plus altitude) can be determined.

The control component is described as "a global network of ground facilities that track the GPS satellites, monitor their transmissions, perform analyses, and send commands and data to the constellation [1]." Figure 1 shows the current operational control component.

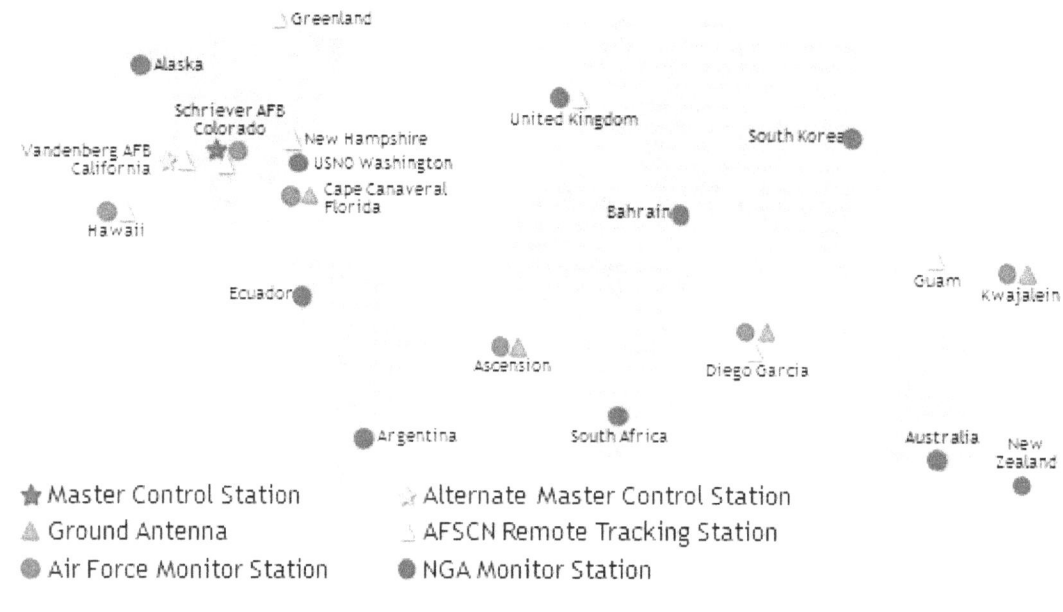

Figure 1. Control Component [1]

The master control station (MCS) and alternate MCS provide command and control of the system. They send and receive navigation messages, assess and preserve health and accuracy of the satellite constellation, and reposition satellites to uphold an optimal GPS constellation. The 16 monitoring stations track and collect atmospheric data, range/carrier measurements, and navigation signals on the satellites and route this information back to the MCS. The four ground antennas transmit command and control to the satellites. The ground antennas are responsible for navigation data and processor program uploads/retrievals, telemetry collection, command transmissions, anomaly resolution, and early orbit support.

Military GPS and Civilian GPS differ in two areas. The first is in terms of enhanced security and jamming resistance. The second area is the type of service. Standard Positioning Service (SPS) is for civilian usage. SPS broadcasts on one frequency and is able to pinpoint an object to within 330 feet. Higher quality SPS receivers with the combination of augmentation systems are able to provide better than 11.5 feet horizontal accuracy. Military uses Precise Positioning Service (PPS), which is encoded and scrambled for security reasons. PPS broadcasts on two frequencies allowing for a reduction in radio degradation allowing for an accuracy within 3 feet. Security and jamming resistance will always be a staple of military GPS, but ongoing modernization of the GPS is reducing the service accuracy gap between civilian and military.

Transmitter power for a satellite is only 50 watts or less. This is a huge factor in the limitation and susceptibility of the system. With this low power atmospheric effect, sky blockage, and receiver quality account for a few of the sources of GPS signal errors [1].

2.3 GPS Spoofing

GPS spoofing uses a signal to take over a GPS receiver. The signal then causes the receiver to deliver false information on the true position of the receiver as dictated by the attacker. Cavaleri remarks that "spoofing is more deceitful" due to the ability of the attack to be unrecognizable which leads to false information being fed to the user [6].

Spoofing is divided into three categories, simple, intermediate, and sophisticated. Figure 2 visually describes the differences between the three categories.

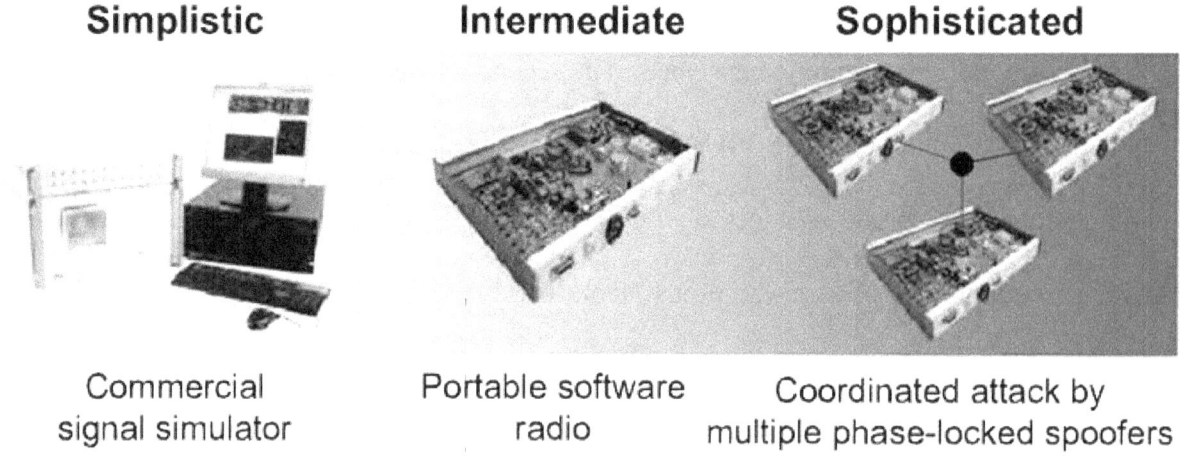

Figure 2. Spoofing Categories [7]

Figure 3 shows the effect of an intermediate attack.

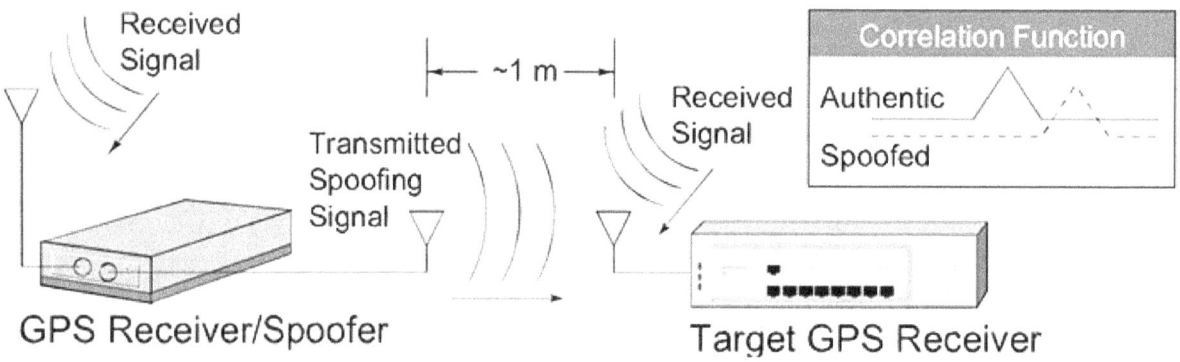

Figure 3. Intermediate Spoofing Attack [7]

Spoofing is effective since the victim is normally unaware of the present attack. This leads to there being no true countermeasures for spoofing because of its unassuming effects. There has been study of an in-line RF device connecting to a GPS antenna, which provides spoofing protection without upgrades to a legacy receiver. Signal quality and autonomous integrity monitoring help in the detection and mitigation of spoofing [5].

Simple and cost effective countermeasures can be retrofitted on nearly all existing GPS receivers, although this will not truly eliminate spoofing attacks [8]. This fact is worrisome, but it also helps in understanding the spoofing attack itself. A successful spoofing attack needs only a target for most assaults. A means of prevention understands the requirements for a successful attack. Locations and precision for an infinite amount of receivers being spoofed to a random location or even a satellite-lock takeover can be identified based on whether the target is civilian or military [9]. This makes it a key variable to study when evaluating GPS attacks.

2.4 GPS Jamming

GPS jamming is the use of a device to block, jam, or interfere with GPS systems communications. The jammer is described as radio frequency transmitters that block or interfere with authorized communications.

Jamming can be done mechanically or electronically. Various mechanical jamming devices are chaffs and corner reflectors. Chaffs and reflectors, which have the same effect come in different shapes, reflect frequencies to produce false navigation solutions.

The more common jammers are electronic which are primarily barrage, spot, and sweep. Barrage jamming sends noise out to multiple receivers at once. While it can affect more targets, the effect is less pronounced due to the increase in spread. Sweep jamming also affects multiple targets, but not simultaneously. It shifts from one target to the next. Spot jamming focuses on only one target. These jamming techniques can be upgraded to other types of jamming such as base jamming, which is barrage jamming at one radar attacking all of its frequencies.

Countermeasures are used to combat jamming. These are more widely known due to the ease and availability of jammers compared to spoofers. Countermeasure techniques are constantly alternating a frequency used, cloaking outgoing signals with random noise, and effective operator training. [1]

GPS signals have lower power signals, making them highly susceptible to interference by intentional and by unintentional means such as radio transmissions [10]. This makes it a key variable to study when evaluating GPS attacks.

2.5 Simulation

Simulation is a useful tool in running multiple tests without sizeable cost, time, or resource impact. As an example, [4] used simulation to understand the effect of spoofing signals by observing the chip delay of a spoofing signal using code and frequency tracking as a response. The steps used for the simulation is to generate an intermediate spoofing signal; the spoofers synch the spoofed signal with the current signal; the GPS signal receives the spoofing signal and then processes it as the correct signal; and the tracking loop error is generated and a "pseudo-range" is established according to the chip delay of the spoofing signal. The navigation solution is then checked to see if it has been adversely affected by the spoofing signal. [4]

One common analysis technique used with simulation-based analysis is Design of Experiments (DOE). Unlike industrial DOE, simulation DOE does not need to use a fractional design; can generate multiple replications without significant cost; there is no need to randomize the runs; and there is greater control over factors within the experiment. [11]

2.6 Regression/Design of Experiments

A common experimental design is a 2^k design; where there are k factors being observed in a low/high setting. The analysis of this resulting design often leads to a regression equation. The equation can allow for better understanding of the design. The equation becomes a predictive model where significant factors and their coefficients are statistically known. [12]

Additionally a response surface can be gleaned from the regression equation. The response surface is a geometric figure of the predictive model. It can show the entire surface of a response based on its factors in a design space (often in two-dimensional views). This can be extremely helpful in optimization as well as in understanding a process. [13]

2.7 SEAS and Related Studies

With its initial release in 1994, System Effectiveness Analysis Simulation (SEAS) provides "a constructive modeling and simulation tool that enables mission-level Military Utility Analysis (MUA)…created to support developmental planning and Pre-Milestone "A" acquisition decisions for military space systems. [14]" It provides agent-based modeling that can be manipulated with programmable rules within a physics based battle environment. It is currently in use by the Space and Missiles System Center, Air Force

Space Command, Air Force Research Laboratory, Air Force Institute of Technology, U.S. Air Force Academy, National Reconnaissance Office, Army Space and Missile Defense Command, and Commander U.S. Pacific Fleet within the DoD. It also is in use by DoD contractors to include Lockheed Martin, The Boeing Company, and Northrop Grumman. [14]

AFIT has done studies using SEAS. Due to the development of computer networks as an integral part of information flow in combat, research is ongoing to gauge the effectiveness and efficiency of this domain. Honaburger [15] uses SEAS to explore network centric warfare (NCW) metrics in military worth analysis. His analysis looks at measures that are import to successful NCW in a Kosovo scenario such as target detection distance outputs, average communication channel message-loading metrics, and target kills to understand better the cognitive domain.

Our study uses SMC/XR's SEAS model [16] to present an urban canyon scenario where PNT is highly stressed. It describes a special operations force (SOF) team trying to recover a weapon of mass destruction (WMD) against a red military force. The team uses GPS to lead themselves from an initial drop point to the WMD and back to an evacuation point while trying to avoid red forces. [16]

2.8 Summary

GPS is a tool that military and civilian organizations use for a myriad of purposes. Beginning with NAVSTAR ran by the USAF, both civilian and militarily share the same system from the control component down to the receivers. Although alike in nearly every area, military application has the advantage of enhanced security and jamming resistance. Since GPS is such an important tool, attacks against users of the system is a given where the most well known are jamming and spoofing. Jamming involves mechanical or electronic means to block or interfere with communications. In spoofing, attackers feed false information to receivers causing positional errors. Spoofing is considered more deceitful since the victim is typically unaware of the attack.

Simulation is a helpful tool in understanding the way a system works without the heavy resource burden that affects other testing means. SEAS is a combat simulation tool that models mission level activities. We briefly describe a SEAS model built by SMC/XR [16] to analyze PNT performance in an urban canyon environment. In Chapter 4, we use this model to explore further the effects of GPS jamming and spoofing for this scenario. Our analysis uses DOE to allow an efficient and effective way to test across an entire operating space. In addition, we develop predictive equations and response surfaces from the simulation outputs retrieved from our experiments to provide a better understanding of how jamming and spoofing effect a PNT system.

III. Methodology

3.1 Introduction

SEAS is used to model the effects of jamming and spoofing on military operations. The ability to repeat and to explore a large number of different scenarios within an experimental design helps to find adequate results; this is a benefit that a simulation provides. SMC/XR models a scenario that uses SEAS, which depicts a recovery mission by a SOF team using GPS [16]. We start with this model and vary selected input parameters to characterize a degradation in PNT performance due to jamming and spoofing effects. A number of output statistics are collected as responses for the detailed study.

3.2 Background

SMC/XR, the SEAS model managers along with SAIC and ExoAnalytic Solutions presented a study at the Military Operations Research Society Symposium (MORSS) in 2009 [16]. The focus area captured in their study is the urban canyon environment. This environment is marred by GPS gaps such as tall buildings and indoor locations. The measures of performance (MOP) used as factors were developed from STRATCOM PNT Joint Capabilities Document and Functional Solutions Analysis and AFSPC Space Force Enhancement FY08 Mission Area Plan [17]. The three MOPs pulled from these documents are availability (assured access to PNT in any condition or environment), accuracy (conformance between a measured and a true PNT parameter), and timeliness (determine if PNT need is met within user defined time parameter) [16].

14

The blue goal in the scenario is to recover a weapon of mass destruction (WMD) from terrorists (red force) located in a target building. The simulation unfolds with SOF team (blue forces) in the target building, recovering the WMD, and maneuvering to the extraction point. PNT is used to navigate and avoid red force and local police. The model behavior is gleaned from the Marine Corps Warfighting Laboratory Instructions, the Army Infantryman Input, and historical operations [18]. Major results from the study are "time lost correlates linearly with the availability of navigation sources, there is a knee in the mission success curve as availability passes 70%, and a distinct floor and shelf indicate that the availability of PNT potentially doubles mission success." [16]

3.3 Measures of Effectiveness

A key piece of constructing the model involved developing measures of effect (MOE) selected from the Joint Staff's Universal Joint Task List (UJTL) [19]. These MOEs are percent of friendly causalities, friendly forces movement delays in hours, and time to seize objectives. For this study, we use multiple MOEs to capture degradation of the PNT system, which are time to mission completion, number of blue team killed, number of engagements, and number of GPS minutes lost.

3.4 Model Behavior

The model focuses on a location in the Middle East between a blue force and a red force. The blue force is made up of Air Force, Navy, and Army units. The red force is made up of the civilians, police, and red military. A screen shot of the model running is shown in Figure 4.

Figure 4. Initial Phase of SEAS Scenario

The Air Force hierarchy, shown in Figure 5, consists of Predators, Global Hawks, and CV-22s. In this scenario, they provide no offensive capability. The remotely piloted aircrafts (RPAs) only deliver sensor and communication ability that highlights and distinguishes red forces. The CV-22 delivers and picks up blue forces with a constant

evacuation time. The Navy unit is a carrier with its mission to deploy and receive the Air

Force's helicopter. [16]

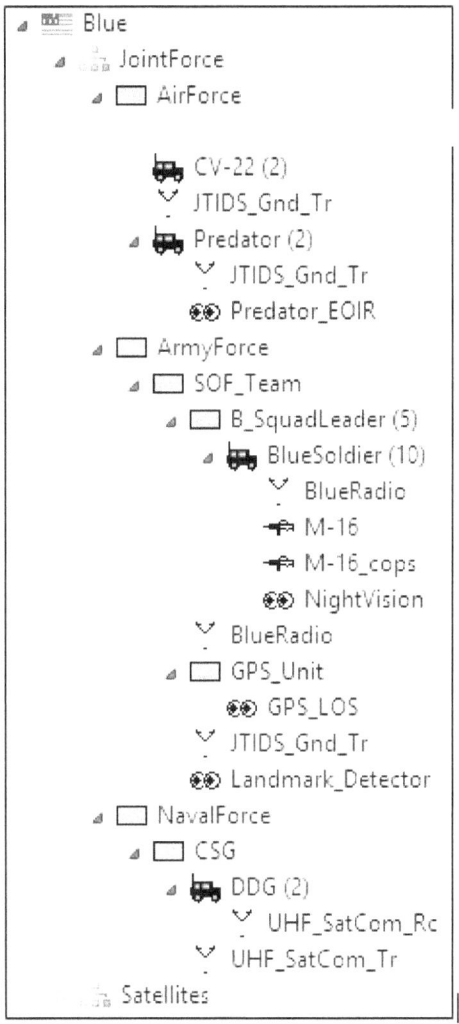

Figure 5. Blue Hierarchy

The Army consists of squad leaders and soldiers. The squad leader is a solider

with the ability to issue commands. There are 50 soldiers in the SOF team with no more

than four groups at one time. The soldier is armed with M-16s, night vision, PNT device, blue tracking, and a radio. A soldier does not engage the enemy unless fired upon, but when engaged, continues engagement until all enemies are killed, enemies flee, or until an order to retreat is given. The blue unit ignores civilians and avoids the police force. The max range of sight for a blue force member is 200 meters. Movement speed is variable and only on foot. The max speed is 5.25 mph. When a soldier is killed, the nearest soldier carries the dead soldier. There is no delay if one soldier is carried, but if two soldiers are being carried then the speed of the soldier is reduced by 25%. [16]

There are 31 GPS units assigned to the blue force. For this scenario, each unit is available. GPS units follow actual unclassified orbits but do not model any signal traffic. The blue GPS unit under SOF team represents a GPS receiver and simply counts the number of satellites visible to the SOF team every time step. PNT system performance and degradation is captured through use of the timeliness, accuracy, and availability. [16]

The red military has six platoons housing multiple soldiers each. Each soldier is armed with a radio, night vision, and a rifle. Trucks move soldiers to a combat area. There are six trucks with 11 soldiers per truck. Forces can join to create larger forces. When retreating, red soldiers retreat as a group, always in the opposite direction of the SOF team. The red soldiers gather information through police chatter, although the groups are not in collusion. There is an initial lag time of 30 minutes from the first police call to when red soldiers move from their base. When traveling in trucks max speed is 45 mph. When traveling on foot max speed is 5.25 mph. Their max sight range is 200 meters. [16]

The police force has six stations each housing multiple police officers. Each officer has a radio and a pistol. There are five police officers per station. Police are on scheduled regular patrols within a vicinity of their assigned police station. They do not engage unless attacked. For this scenario, if fired upon, the police run away. Their max speed on foot is 5 mph, on motorcycle is 50 mph, and in station wagon is 50 mph. Their max sight range is 50 meters. [16]

The civilians are non-offensive units that are only located downtown. At the start of the mission, there are 20 civilians on the street. A random counter adds more civilians the longer the campaign goes along, where 20 more civilians are added each hour. If engaged, civilians run in random directions. There is a 50% chance they call the police, where the location information takes 5 minutes to enter police chatter. Tactical Programming Language (TPL) for this logic is shown in Figure 6.

```
!! Randomly move (test--until we have road network)
rdist = 0.25*uniform
rdir = 360*uniform
Move me->           + rdist*vector rdir
While me->          != me->
    If me->           > 0 & Called911 == 0
        If uniform    > 0.5
            me->"CellPhone_Tr"->          = 1    !! 50% chance to report Blue to Police
            Call911 = me->
            Broadcast Call911
            Call911 = 0
            Called911 = 1
            me->           = 0
        Else
            me->"CellPhone_Tr"->          = 0
        EndIf
    EndIf
EndWhile
If GC_Dist me->          , Downtown  > 0.5
    Move Downtown      !! move toward center if too far
    Delay 5            !! give it time to move before new move given
EndIf
```

Figure 6. Civilian Protocol

The scenario follows the outline shown in Figure 7 and discussed below. The blue forces are deployed to the drop off location. The special operations forces (SOF) team then makes its way to the building housing the WMD. The SOF team clears the building and recovers the WMD. No SOF team or enemy forces will enter a building once the WMD is recovered. WMD recovery begins the evacuation phase.

3.5 Engagement Criteria

There are three unique engagement criteria during the scenario. During the "Enemy Fights" criteria, the red soldiers always engage the SOF team. During an attack, if needed the red units wait for reinforcements. If at any point the enemy military force does not outnumber the SOF team at least 3 to 1, then the red soldiers disengage and retreat. This engagement criterion is the "Enemy Runs Away" and is triggered by the

SOF team killing at least 50% of the red soldiers. The enemy forces always shadow the SOF team until reinforcements have arrived. The last criterion is when the SOF team retreats. This occurs when the red force kills greater than 50% of the SOF team. The SOF team never fully leaves as it still keeps it mission of recovering and delivering the WMD.

3.6 Evacuation Criteria

The evacuation point is where the CV-22 picks up and recovers the WMD and teams. At this location, there can be no more than 3 to 1 red units on location or there can be no pick up. The CV-22 could come under heavy fire and evacuation will not be possible. The CV-22 evacuation delay time is constant and its PNT is not degraded. There are two defined evacuation points in the model where the SOF team moves towards the closest from route taken which varies with engagements and other factors.

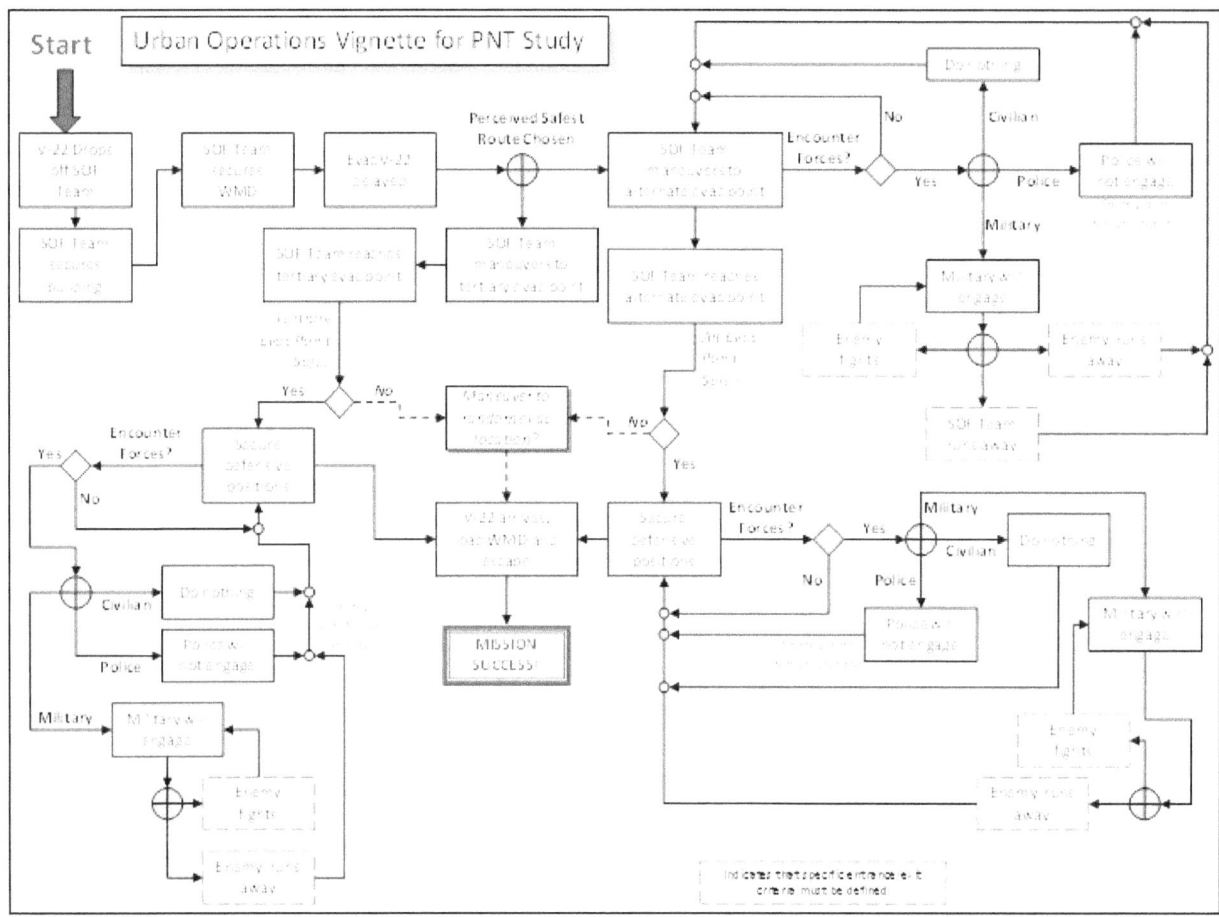

Figure 7. Urban Operations Vignette [16]

3.7 Selected Measures

For the purpose of this model, the three inputs are availability and accuracy, which are measures for a jamming capability, and timeliness, which is a measure for spoofing. This study looks at these variables to assess what jamming and spoofing does to mission success. Does jamming have a bigger effect on mission success or is spoofing the more problematic attack? What is the response depending on the overall modeled area for jamming and spoofing? These questions are what the study looks to answer.

Although availability and accuracy are grouped to describe the jamming capability, all three variables are systematically changed during the experiment.

The inputs are continuous variables. Accuracy is conformance between a measured and a true PNT parameter, where the position error measured in meters, ranging from one to 100, where one is the most accurate. Availability is the assured access to PNT in any condition or environment ranging from zero to one, where one is fully available. Timeliness, the time to recover a lost signal, is measured in seconds, ranging from five to 360, where five is the best timeliness for our scenario.

Responses are mission success, time to mission success, number of blue team killed, number of engagements, and amount of time with no satellite GPS signal. The responses are discrete or continuous. Time to mission success is a continuous response measured in minutes, with the determination of it being accepted for analysis based on if there is a mission success. Number of blue team killed and number of engagements are discrete values. Amount of time with no satellite GPS signal is a continuous response in minutes.

3.8 Constant Measures

The other parameters varied in the SMC study [16] are held constant for our study to focus on effects of jamming and spoofing at different levels for our selected responses. For our scenario there are two predators used to help the blue forces identify red forces. The police do not join the fight. There is a 3-minute delay by blue forces when initially dropped off at the starting location.

23

3.9 Analysis Approach

To understand fully any effect of each factor, a DOE is employed. The DOE is a 5^k design, where the response is measured at combination of unique levels for each of the k factors. An empirical model is created and tested for model parameter significance by p-values while confirming assumptions such as constant variance and normality. A response surface is then made comparing variables to one or multiple responses.

3.10 Verification and Validation

Confirmatory runs are used to verify that the predictions received from the experiment are valid. Many times an estimated optimal setting is used to verify whether the response surface or prediction formula meets the goal of the experiment. Since this is a simulation, the environment of the experiment is the same and at least three runs are normally sufficient. In addition, consultation with subject matter experts (SMEs) and comparison to past data are used for verification and validation.

3.11 Summary

We describe an SMC/XR scenario built with SEAS where PNT is highly stressed as a SOF team recovers a WMD from a red force using GPS. Our study uses this scenario to understand and capture the degradation of the PNT system due to spoofing and jamming. We model spoofing and jamming by varying availability, accuracy, and timeliness parameters as factors in our DOE to characterize the system. The scenario uses a blue force made up of Air Force, Navy, and Army units against a red force comprised of civilians, police, and a red military force. GPS units used by blue forces in the model do not use actual GPS signals, but count the number of satellites visible to the SOF team

every time step. The DOE discussed in Chapter 4 produces the data used to create a

prediction model and response surfaces for comparing variables effect on responses. An

in-depth look at two replications provides additional insight and serves as an additional

verification and validation tool.

IV. Analysis

4.1 Overview

This chapter outlines the analysis approach used and provides the results of the DOE. The chapter examines four responses: time to complete the mission, SOF team members killed, number of engagements, and the number of minutes the GPS signal is lost. Where applicable, confirmatory runs and response surfaces are used to validate the prediction models. This chapter also compares the two jamming components, availability, and accuracy, to the spoofing component of timeliness.

4.2 Setting up the DOE

Initially a setup of the variable space is needed to focus the experiment. The goal of the experiment is to see meaningful change in the design. Setting all of the variables at high points would not produce a meaningful design, as there would not be a better understanding of the reduction in PNT. Likewise setting all of the variables at the low points would only serve to show what going into a combat zone blind would entail. Since GPS has become integrated in many aspects of both civilian and military products, setting all variables at the low point would not provide much knowledge for this experiment. A 10% reduction is taken from the high and low points to show what the variation in PNT produces. To understand better this variation, replication runs are included in the experiment. The more test points collected is always better. Even though there is not a monetary expense associated with this simulation, time is a concern. On average for this simulation, the SEAS simulation tool takes an hour to do 50 runs, with runs experiencing heavier losses taking longer. We set a budget of 1000 runs for the study. To balance the amount of design points with replications, the level of the DOE was considered. A 4^3 full factorial model requires 64 runs, while a 5^3 full factorial model requires 125 runs. Increasing the number of levels gives a better understanding of the complete design space. To maximize time and design points a decision between a 4^3 with 15 replications or a 5^3 with 8 replications is made. Ultimately, the ability to understand better the design space is the reason that the 5^3 design is chosen.

4.3 Design

Table 1 shows the 5^3 design space in engineering units along with the coded units. The coded space maps the engineering units to the range -1 to 1. This yields orthogonal columns, which means all the elements in the column sum to zero.

Table 1. Design Space

Accuracy	Availability	Timeliness
Engineering Values		
10	0.1	36
30	0.3	108
50	0.5	180
70	0.7	252
90	0.9	324
Coded Space		
1	-1	1
0.5	-0.5	0.5
0	0	0
-0.5	0.5	-0.5
-1	1	-1

27

4.4 Results Introduction

After running the simulation, the experimental results contain 1000 runs, which is a five level DOE with three factors and eight replications. Eighty of these runs caused erratic results in the simulation. Low availability appears to be the cause of distress. These resulted in high completion times and abnormally high minutes lost for the GPS signal. These runs were removed since the results do not give accurate readings at the affected levels. Overall, 46 design points are affected. These design points have between three and seven replications. The remaining 920 runs make up the data analyzed.

The average and standard deviation rates realized by each factor, accuracy, availability, and timeliness, at each level are recorded in Table 2, Table 3, and Table 4, respectively.

Table 2. Average and Standard Deviation Accuracy Results

	Accuracy				
	10	**30**	**50**	**70**	**90**
Time	262.29 ± 170.90	382.53 ± 227.73	388.977 ± 234.33	401.81 ± 248.95	401.76 ± 227.57
Blue Killed	32.82 ± 6.07	33.41 ± 5.98	32.07 ± 5.63	32.63 ± 5.94	33.67 ± 5.87
Number of Engagements	11.41 ± 2.24	11.80 1.88	11.91 ± 1.81	12.01 ± 1.92	11.92 ± 1.70
Minutes GPS Signal Lost	173.20 ± 167.06	295.26 ± 225.62	305.28 ± 230.88	318.85 ± 247.98	317.49 ± 227.56

Table 3. Average and Standard Deviation Availability Results

	Availability				
	0.1	**0.3**	**0.5**	**0.7**	**0.9**
Time	393.10 ± 238.26	389.61 ± 232.47	375.98 ± 228.94	345.72 ± 227.57	324.51 ± 210.04
Blue Killed	32.87 ± 5.63	33.32 ± 6.12	32.66 ± 5.56	32.91 ± 6.35	32.83 ± 5.93
Number of Engagements	11.80 ± 1.71	11.98 ± 1.91	11.70 ± 1.87	11.79 ±2.21	11.74 ± 1.93
Minutes GPS Signal Lost	308.16 ± 235.80	304.07 ± 230.67	288.72 ± 228.10	259.41 ± 226.77	242.06 ± 210.06

Table 4. Average and Standard Deviation Timeliness Results

	Timeliness				
	36	**108**	**180**	**252**	**324**
Time	171.35 ± 91.14	300.49 ± 190.58	380.09 ± 206.35	471.91 ± 207.50	554.30 ± 219.44
Blue Killed	33.00 ± 6.00	33.36 ± 6.18	32.91 ± 5.91	32.39 ± 5.86	32.83 ± 5.57
Number of Engagements	10.97 ± 2.10	11.57 ± 1.99	12.02 ± 1.71	12.38 ± 1.70	12.25 ± 1.74
Minutes GPS Signal Lost	72.00 ± 69.08	216.17 ± 180.22	297.96 ± 200.80	392.13 ± 205.06	472.85 ± 218.92

Based on these results, an initial conjecture is made concerning the effect of individual factors on each response. Looking at the Accuracy factor, the time the SOF team runs the mission and the average minutes that GPS signal is lost appear to be affected by the accuracy factor. For the Availability factor, time and minutes GPS signal lost might again appear to be affected. Time and minutes GPS signal lost also appear to be affected by the timeliness factor.

29

Although the tables provide a general understanding of the effect of individual factors on each response, the standard deviation on for each measurement is large. This is due to the wide range for each measurement. One level of a factor is also accounting for all levels of the other two factors. To get a better view, confidence intervals of each factor split by the other factors are plotted against the two responses that appear to be affected, time and minutes GPS signal lost in Figure 8 - Figure 13. All confidence intervals are constructed independently at the 95% level. In each figure, the best settings for all factors are in the lower right corner and worst settings in upper left corner. Looking at the Figure 8 - Figure 10, a high level Timeliness generates a lower time no matter what setting the other factors are set at. Such a blanket statement cannot be said about the other two factors. The same holds true for Figure 11 - Figure 13, where a high level of timeliness produces a lower amount of lost GPS signal minutes, no matter the settings of the other two factors. Since timeliness corresponds directly to spoofing in our study, these initial results indicate a larger impact from spoofing than jamming. To understand fully the effect of each factor a regression analysis and a response surface analysis is conducted.

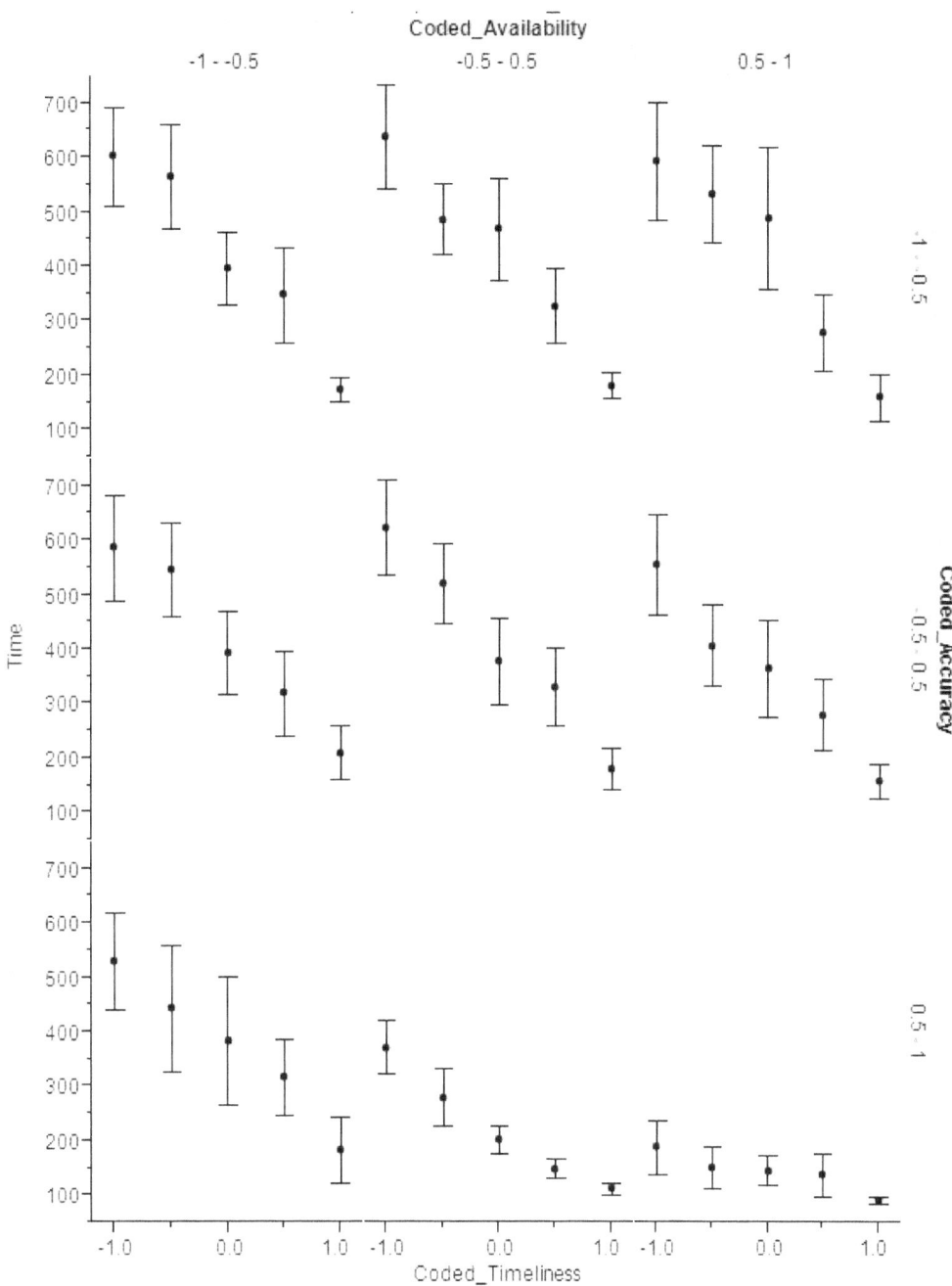

Figure 8. Time Confidence Interval (Timeliness Base)

Figure 8 shows much less effect on timeliness when both availability and accuracy are at the highest levels. With any degradation in availability or accuracy, timeliness shows significant increases going from low to high levels.

31

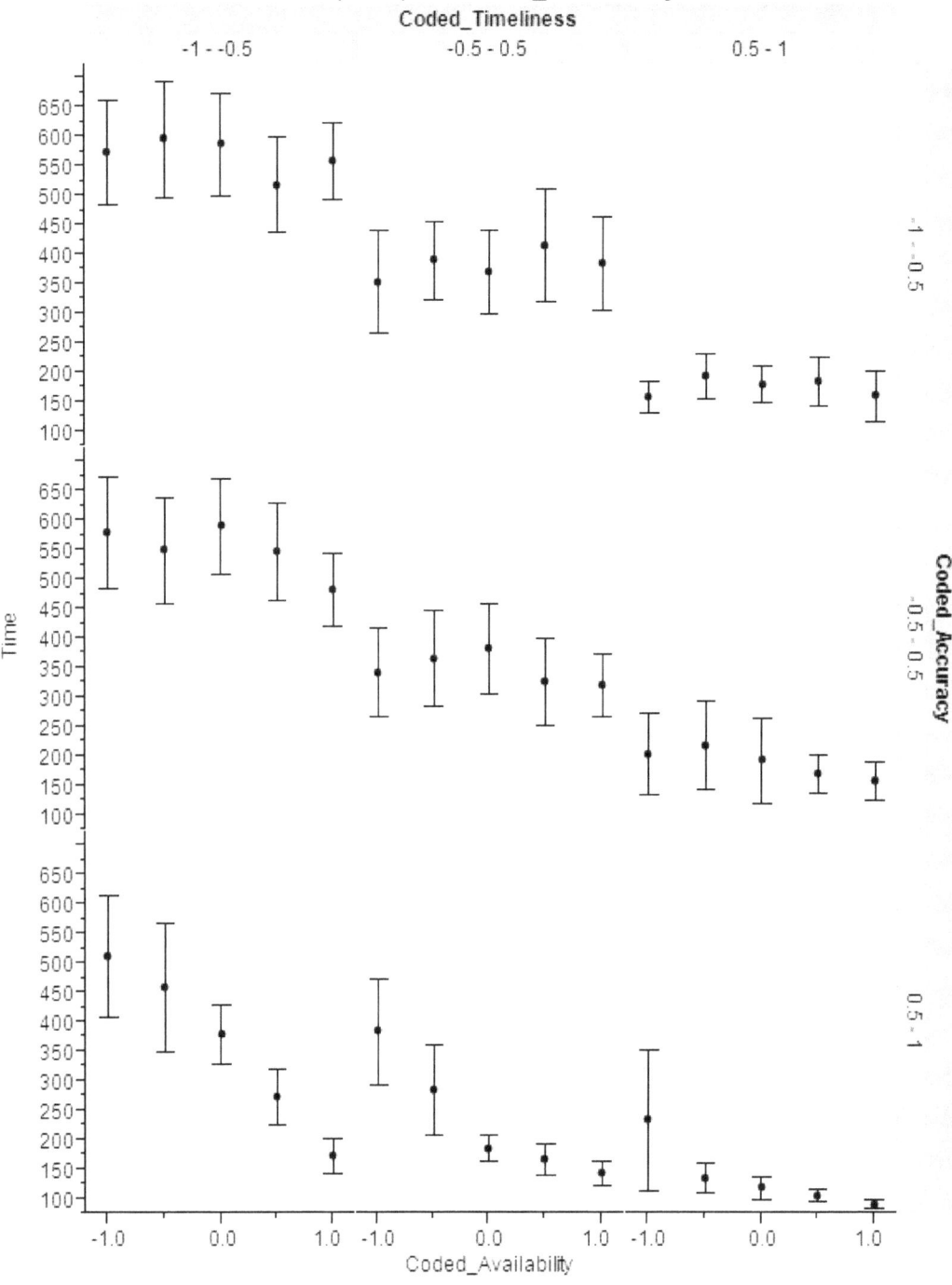

Figure 9. Time Confidence Interval (Availability Base)

In Figure 9, we see the most effect across levels of availability when accuracy is at the best level and timeliness at the worst.

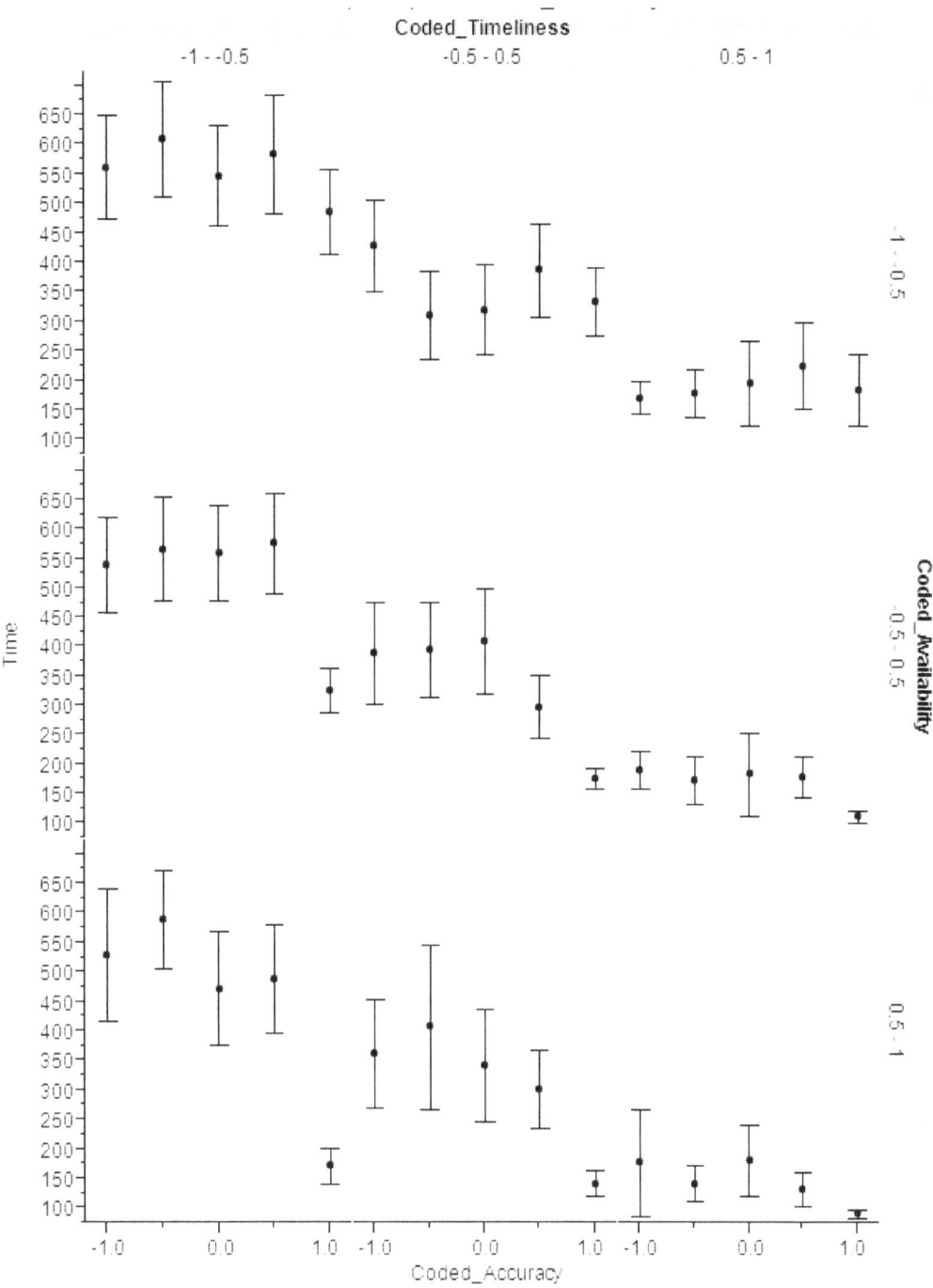

Figure 10. Time Confidence Interval (Accuracy Base)

Figure 10 shows a significant increase in time with an accuracy increase from 10 – 30 meters at degraded levels of timeliness and availability. There also seems to be a significant drop in time when accuracy is at its highest level and timeliness is greater than 108 seconds.

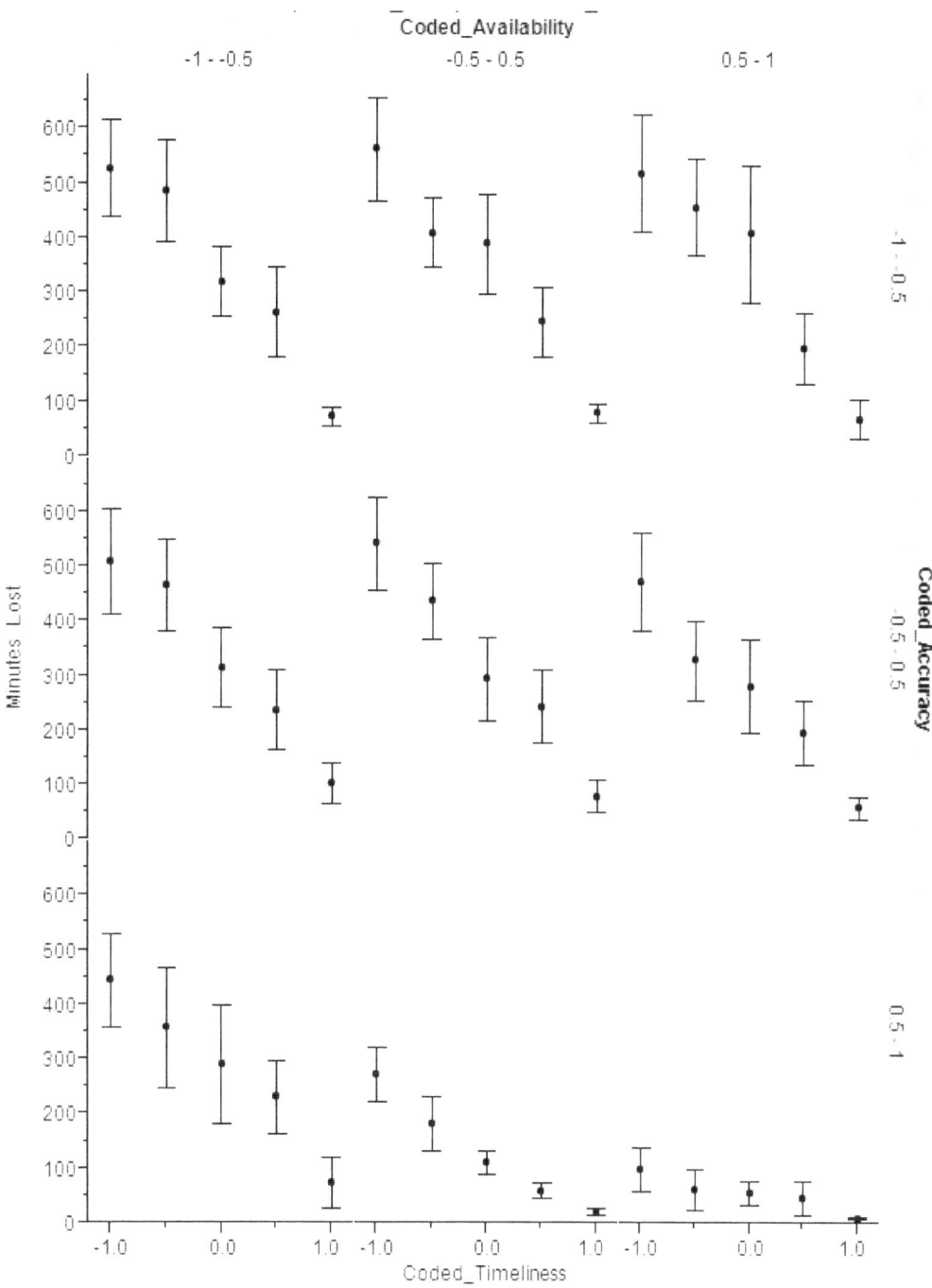

Figure 11. GPS Signal Minutes Lost Confidence Interval (Timeliness Base)

Just as in Figure 8, there is little effect across levels of timeliness in Figure 11 at

the best setting for availability and accuracy. With any degradation in availability or accuracy, we see a significant increase in GPS signal minutes lost with increases in timeliness.

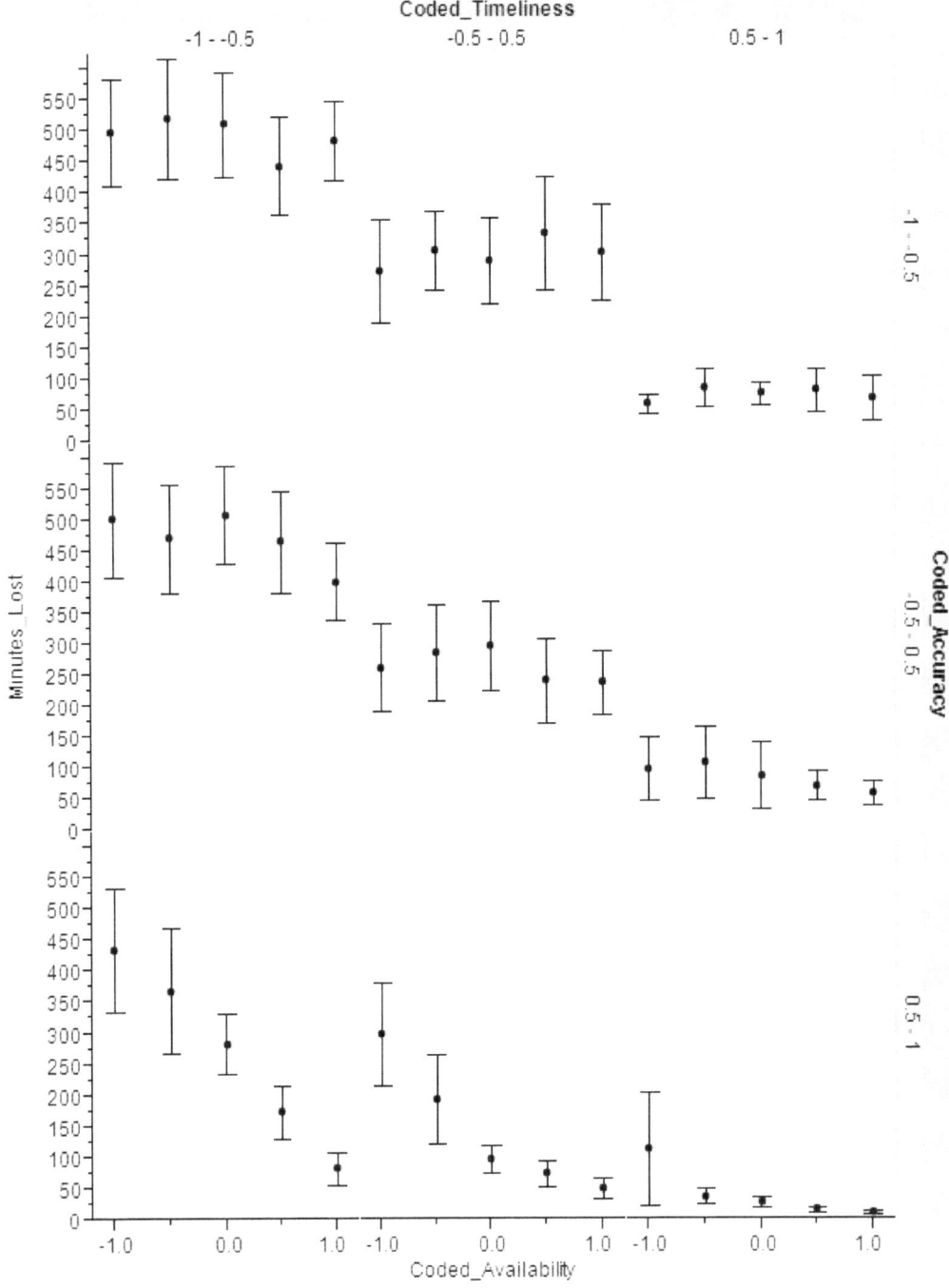

Figure 12. GPS Signal Minutes Lost Confidence Interval (Availability Base)

Figure 12 shows little effect across levels of availability at the best setting for timeliness and accuracy. With any degradation in timeliness or accuracy, we see a significant increase in GPS signal minutes lost with increases in timeliness.

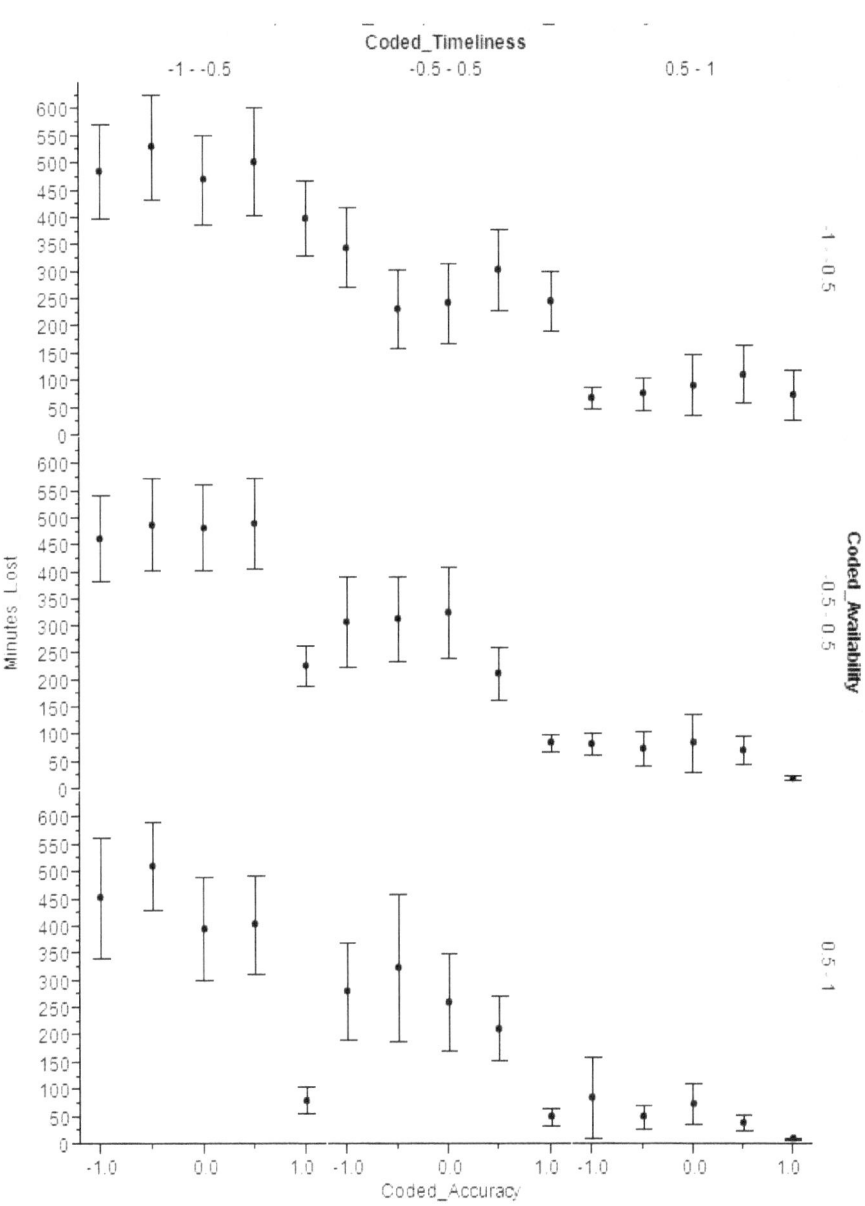

Figure 13. GPS Signal Minutes Lost Confidence Interval (Accuracy Base)

Figure 13 results show a significant increase in GPS signal minutes lost with an accuracy increase from 10 – 30 meters at degraded levels of timeliness and availability. There also seems to be a significant drop in GPS signal minutes lost when accuracy is at its highest level and timeliness is greater than 108 seconds.

4.5 Time Response

We initially construct a model with time as the response for the full factorial design. Looking at the residual by predicted plot in Figure 14, a transformation made sense since the plot took on a funneling shape and not a random even spread across a median. A Box-Cox Transformations analysis shown in Figure 15, is performed and a value of -0.2 is pulled from the analysis. This value being close to zero is interpreted as a Log transformation on the response variable.

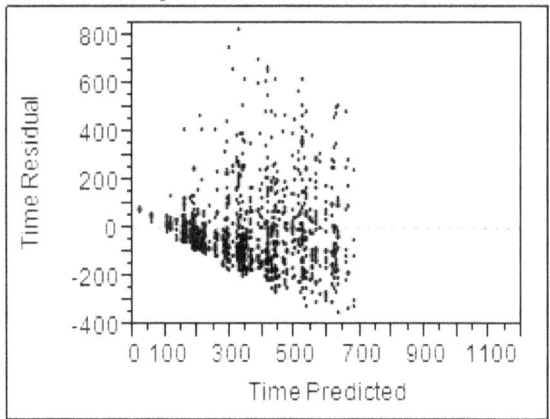

Figure 14. Initial Time Residual by Predicted Plot

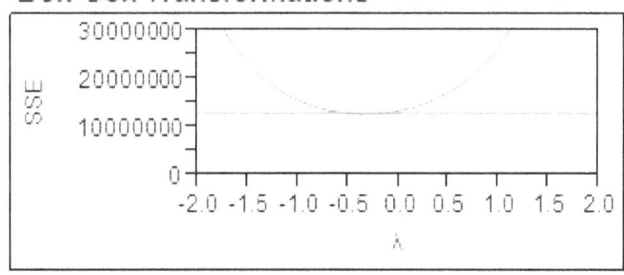

Figure 15. Initial Time Box-Cox Transformations Analysis

Using the transformation and stepwise analysis, we first notice potential signs for higher order terms with the Actual by Predicted plot in Figure 16. The predicted model, shown in Figure 17, contains higher order terms, as indicated from Figure 16. The lack of fit test along with the response surface shows the model contains higher order terms. Model adequacy checking indicates that a constant variance and normality assumption are both reasonable.

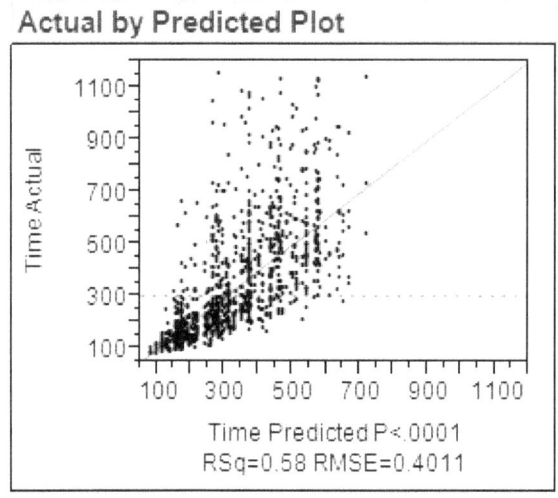

Figure 16. Log (Time) Actual by Predicted Plot

Parameter Estimates

Term	Estimate	Std Error	t Ratio	Prob>|t|
Intercept	5.9259288	0.023832	248.65	<.0001*
Coded_Accuracy	0.0563198	0.056591	1.00	0.3199
Coded_Availability	-0.126877	0.018694	-6.79	<.0001*
Coded_Accuracy*Coded_Availability	-0.236812	0.026231	-9.03	<.0001*
Coded_Timeliness	-0.478252	0.055171	-8.67	<.0001*
Coded_Accuracy*Coded_Accuracy*Coded_Accuracy	-0.291838	0.062629	-4.66	<.0001*
Coded_Accuracy*Coded_Accuracy*Coded_Accuracy*Coded_Accuracy	-0.191567	0.028085	-6.82	<.0001*
Coded_Timeliness*Coded_Timeliness	-0.178668	0.032019	-5.58	<.0001*
Coded_Timeliness*Coded_Timeliness*Coded_Timeliness	-0.125074	0.061628	-2.03	0.0427*

Figure 17. Log (Time) Parameter Estimates

The goal is to minimize the amount of time the SOF team needs to complete the mission, so referring to Figure 18, setting all variables to their highest levels (Accuracy, Availability, Timeliness) = (10, 0.9, 36) results in a time within the interval {70.3718, 85.8437} 95% of the time.

40

Prediction Profiler

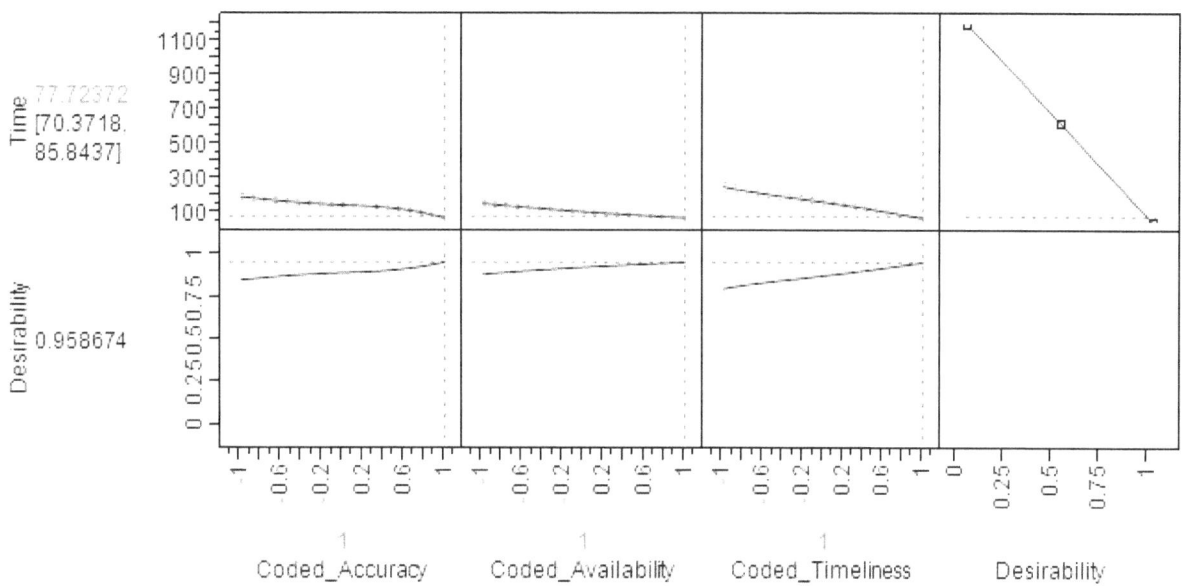

Figure 18. Time Prediction Profiler

To validate this response five validation runs are made, with all factors at their

high levels for all runs. Table 5 lists the results. Across the entire run, the range is 82.95

to 88.55 with an overall average of 85.44. This lines up with the prediction model.

Notice also in Figure 18, we see a slight decrease in mission time for improvements in all

three factors.

Table 5. Time Validation Runs

Run	Time	Average
1	88.55	
2	82.95	
3	85.6	85.44
4	83.2	
5	86.9	

A response surface helps to understand better how the variables influence the response. Figure 19 is the response surface for Accuracy and Availability using the Time response variable. The surface indicates a stationary ridge. This is great news for understanding the model, as there is a line maximum of 431 for the design space of accuracy and availability. In essence, there is little change in the response as we move along the line maximum. The minimum point for this design space is 123 at the maximum levels for both variables. Figure 20 and Figure 21 represent the response surfaces for accuracy and timeliness as well as timeliness and availability. These two are both rising ridges, where there is essentially a rising slope within the design space. For both models, the maximum value is at the low levels for both design variables and the high design variables produce the minimum value being 83 and 113 minutes, respectively.

Surface Plot

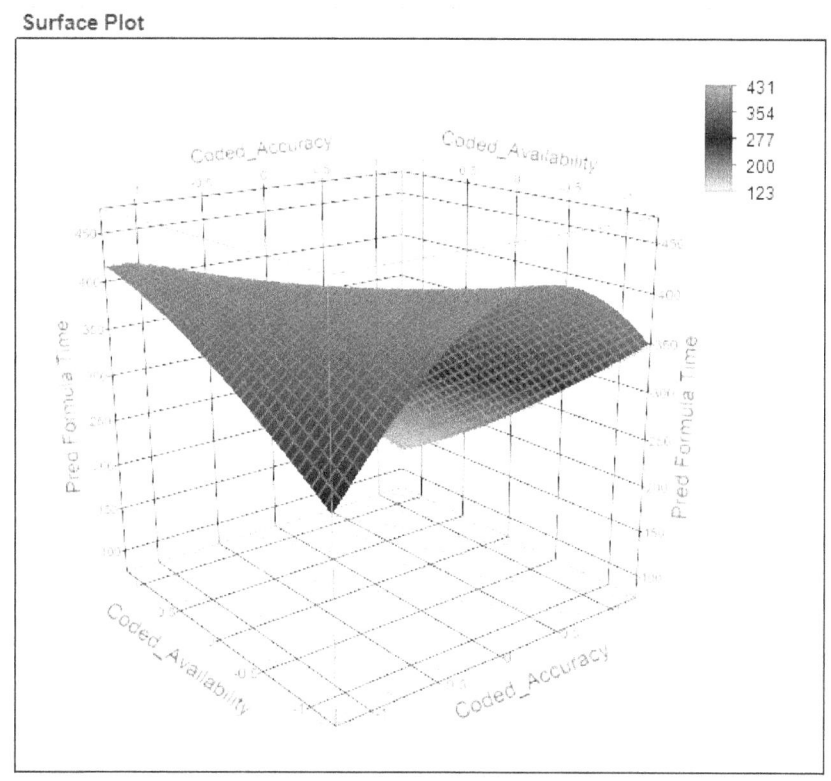

Figure 19. Time Accuracy vs Availability Response Surface

Surface Plot

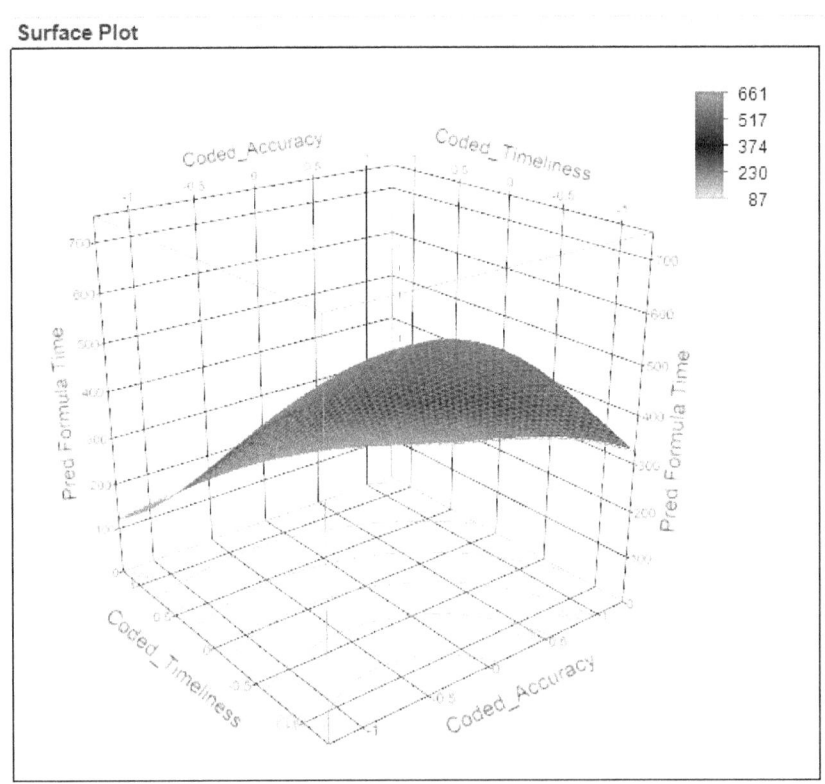

Figure 20. Time Accuracy vs Timeliness Response Surface

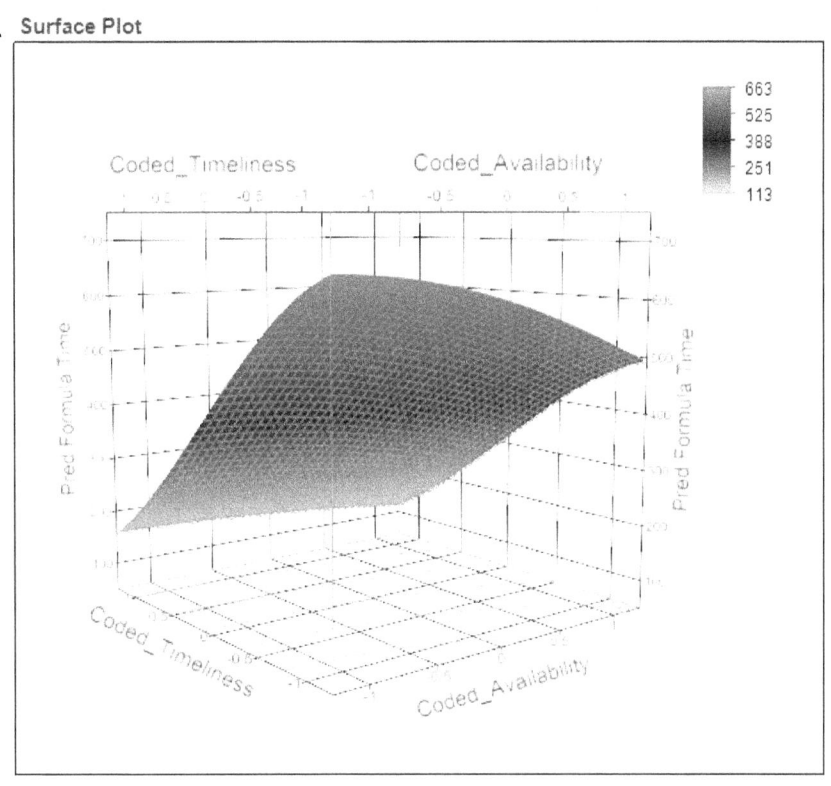

Figure 21. Time Timeliness vs Availability Response Surface

4.6 Blue Team Killed Response

When examining the blue killed response, no "good" model is apparent. After going through all model types, the best model in terms of R^2, ANOVA p-value, and parameter fits is Figure 22. Looking at the Actual by Predicted Plot, the values appear in vertical line. Also looking at the averages of blue killed by factors, Table 2 – Table 4, the factors look have little to no effect. Statistically, PNT has no effect on the number of SOF team killed during the mission.

Summary of Fit

RSquare	0.006945
RSquare Adj	0.003692
Root Mean Square Error	5.906943
Mean of Response	32.9163
Observations (or Sum Wgts)	920

Figure 22. Blue Killed Summary of Fit

Parameter Estimates

Term	Estimate	Std Error	t Ratio	Prob>|t|
Intercept	32.516253	0.304851	106.66	<.0001*
Coded_Accuracy	1.1968306	0.833052	1.44	0.1511
Coded_Accuracy*Coded_Accuracy	0.8043238	0.464529	1.73	0.0837
Coded_Accuracy*Coded_Accuracy*Coded_Accuracy	-1.626698	0.921834	-1.76	0.0780

Figure 23. Blue Killed Parameter Estimates

4.7 Number of Engagements Response

Examining the number of engagements, Figure 25 provides the predicted model. The model contains only accuracy and timeliness, as availability was not a significant factor in the modeling of the number of engagements. Referring to Figure 24, the R^2 as well as the R^2_{adj} are extremely low. This indicates a poor fit of the actual data to the predicted model. No further analysis for this response is conducted.

Summary of Fit

RSquare	0.080558
RSquare Adj	0.077547
Root Mean Square Error	1.855501
Mean of Response	11.80217
Observations (or Sum Wgts)	920

Figure 24. Number of Engagements Summary of Fit

Parameter Estimates

Term	Estimate	Std Error	t Ratio	Prob>|t|
Intercept	12.058535	0.095148	126.73	<.0001*
Coded_Accuracy	-0.270206	0.086151	-3.14	0.0018*
Coded_Timeliness	-0.681151	0.087491	-7.79	<.0001*
Coded_Timeliness*Coded_Timeliness	-0.43228	0.147825	-2.92	0.0035*

Figure 25. Number of Engagements Parameter Estimates

4.8 Minutes GPS Signal Lost Response

We initially construct a model with GPS signal minutes lost as the response for our full factorial design. Looking at the residual by predicted plot, Figure 26, a transformation made sense since the plot took on a funneling shape and not a random spread across a median. A Box-Cox Transformations analysis, Figure 27, is performed and a value of zero is recommended. This value is interpreted as a Log transformation on the response variable.

47

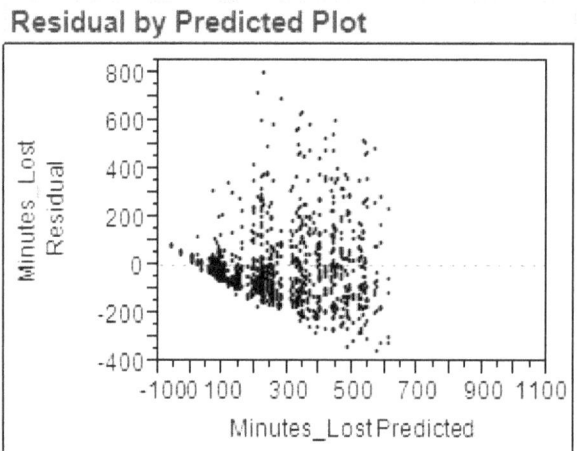

Figure 26. Initial GPS Minutes Lost Residual by Predicted Plot

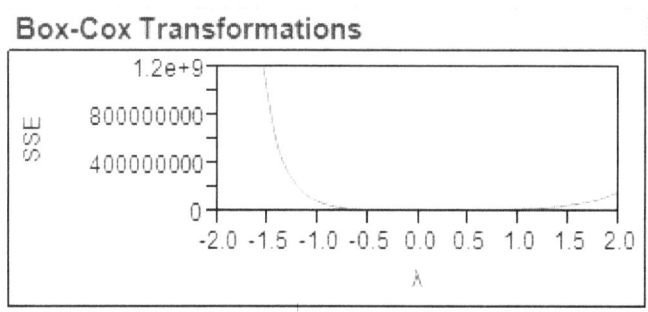

Figure 27. Initial GPS Minutes Lost Box-Cox Transformations Analysis

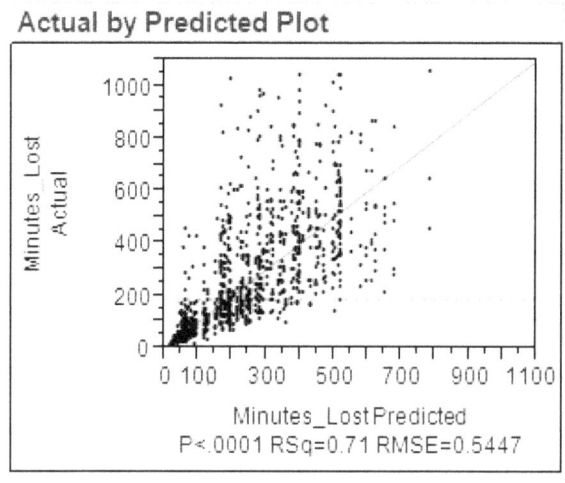

Figure 28. Log (GPS Minutes Lost) Actual by Predicted Plot

48

Parameter Estimates

Term	Estimate	Std Error	t Ratio	Prob>|t|
Intercept	5.644501	0.029118	193.85	<.0001*
Coded_Accuracy	0.1000669	0.076854	1.30	0.1932
Coded_Availability	-0.23291	0.025386	-9.17	<.0001*
Coded_Accuracy*Coded_Availability	-0.439718	0.035626	-12.34	<.0001*
Coded_Timeliness	-0.615332	0.074956	-8.21	<.0001*
Coded_Accuracy*Coded_Accuracy*Coded_Accuracy	-0.538191	0.085052	-6.33	<.0001*
Coded_Accuracy*Coded_Accuracy*Coded_Accuracy*Coded_Accuracy	-0.371814	0.038141	-9.75	<.0001*
Coded_Timeliness*Coded_Timeliness*Coded_Timeliness	-0.439762	0.083741	-5.25	<.0001*
Coded_Timeliness*Coded_Timeliness*Coded_Timeliness*Coded_Timeliness	-0.454748	0.038576	-11.79	<.0001*

Figure 29. Log (GPS Minutes Lost) Parameter Estimates

Using the transformation and stepwise analysis, we first notice potential signs for higher order terms with the Actual by Predicted plot in Figure 28. The predicted model, provided in Figure 29, contains higher order terms, as indicated in Figure 28. The lack of fit test along with the response surface indicates the model contains higher order terms. Model adequacy checking indicates that a constant variance and normality assumption are both reasonable.

The goal is to minimize the amount of GPS Minutes Lost during the SOF team's mission. Referring to Figure 30, setting all variables to their highest levels (Accuracy, Availability, Timeliness) = (10, 0.9, 36) results in a minutes lost within the interval {12.3917, 16.2348} 95% of the time.

Prediction Profiler

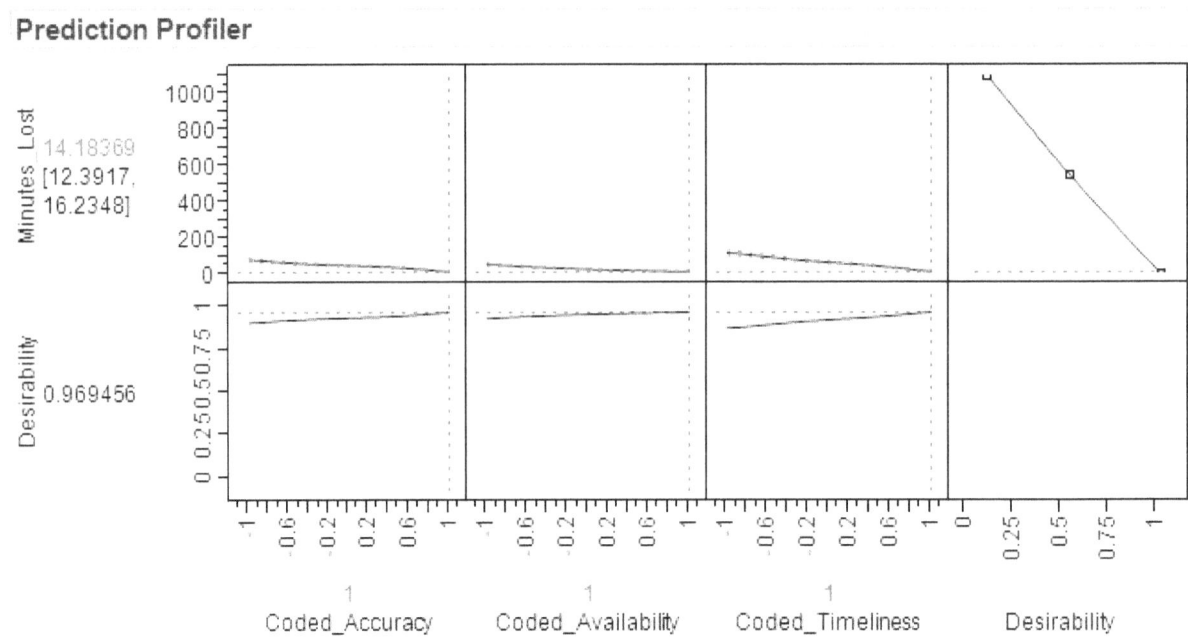

Figure 30. GPS Minutes Lost Prediction Profiler

To validate this response five validation runs are made, with all factors at their
high levels for all runs. Table 6 lists the results. Across the entire run, the range is 10.85
to 14.5 with an overall average of 12.48. Although the three lowest runs are below the
predicted range, the average falls within the prediction model.

Table 6. GPS Minutes Lost Validation Runs

Run	Time	Average
1	11.75	
2	14.5	
3	13.4	12.48
4	10.85	
5	11.9	

A response surface helps to understand better the variables effect on the response. Figure 31 is the response surface of Accuracy and Availability for the GPS Minutes Lost response variable. The surface forms a saddle point. The minimum point for this design space is 38 at the maximum levels for both variables and the maximum point is at 401 minutes lost. Figure 32 and Figure 33 represent the response surfaces for timeliness and accuracy as well as timeliness and availability, respectively. Both surfaces indicate a maximum point within the design space. In Figure 32, the maximum value is within the Accuracy range {45, 85} and within the Timeliness range {252, 324} producing, a value of 555 minutes lost, and the high design variables produce the minimum value being 14 minutes lost. In Figure 33, the maximum value is within the Availability range {0.1, 0.55} and within the Timeliness range {230, 324} producing, a value of 582 minutes lost, and the high design variables produce the minimum value being 26 minutes lost.

Surface Plot

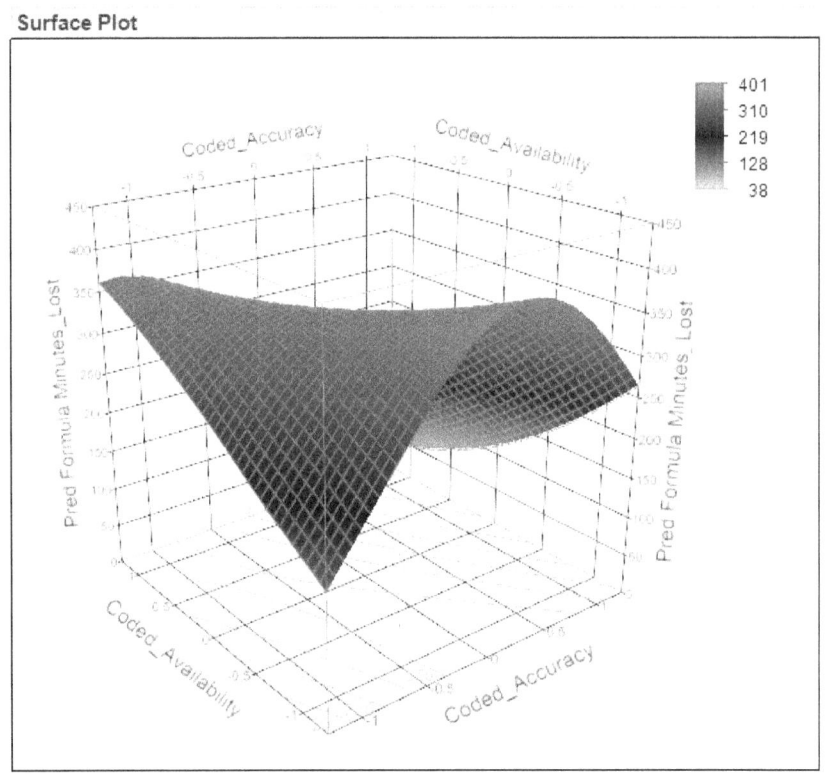

Figure 31. GPS Signal Minutes Lost Accuracy vs Availability Response Surface

Surface Plot

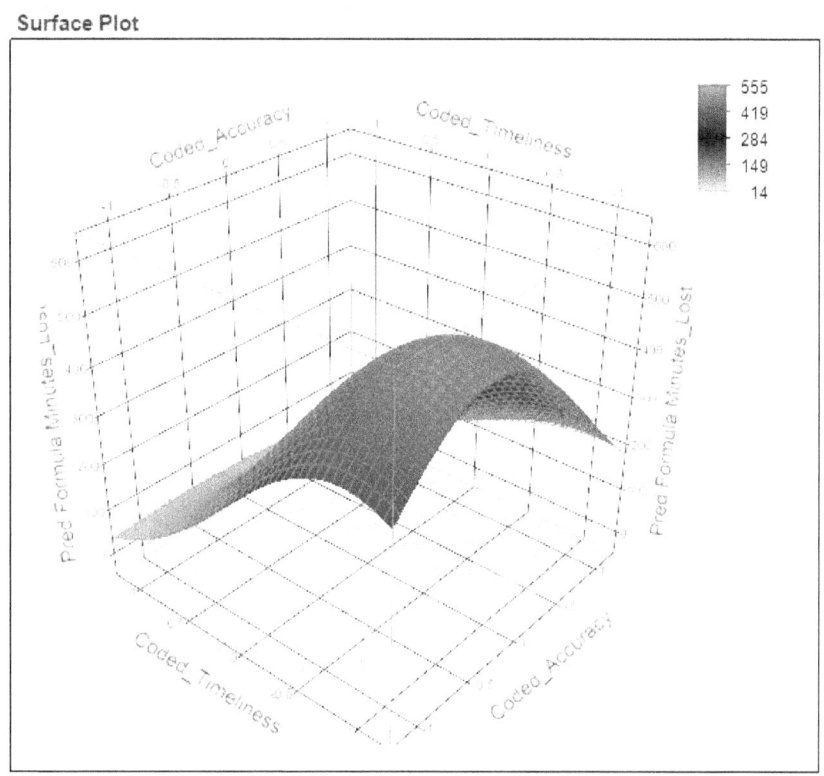

Figure 32. GPS Signal Minutes Lost Accuracy vs Timeliness Response Surface

53

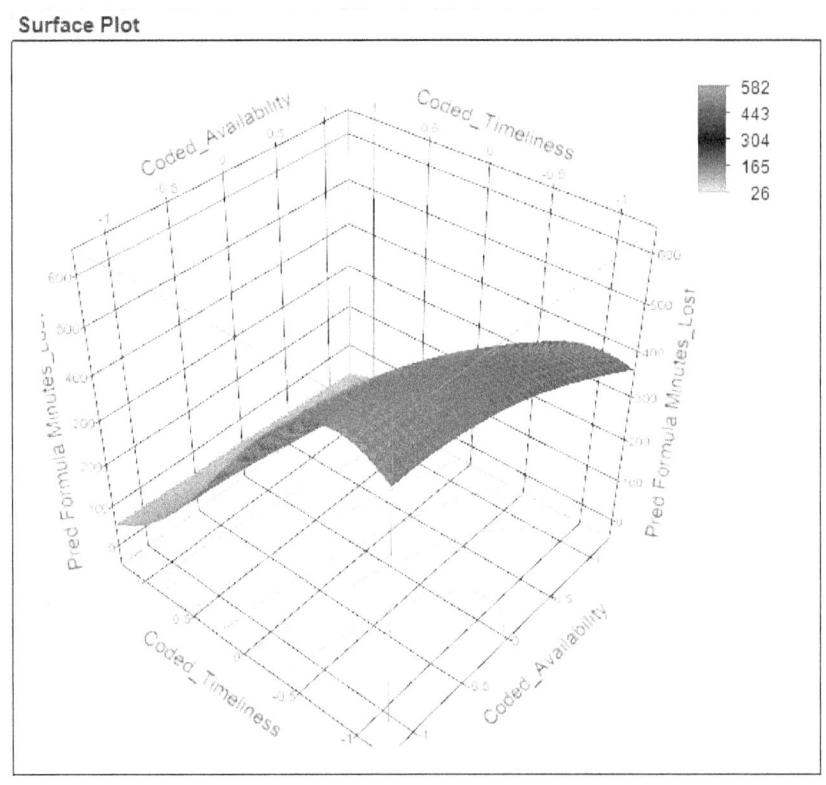

Figure 33. GPS Signal Minutes Lost Timeliness vs Availability Response Surface

4.9 Individual Run

We decided to look at two additional individual runs to gain more insight and to

validate further. The first run depicted in Figure 34, is a mission totaling 72.8 minutes

with only two engagements, 15 blue units killed, and 32.45 minutes from lost GPS signal.

For this run, the factors are high (1) for accuracy, medium-high (0.5) for availability, and

medium-high (0.5) for timeliness. The SOF team recovered successfully the WMD

before being detected and engaged. During the run, PNT issues did not occur until after

recovering of the WMD, but lasted until the end of the mission.

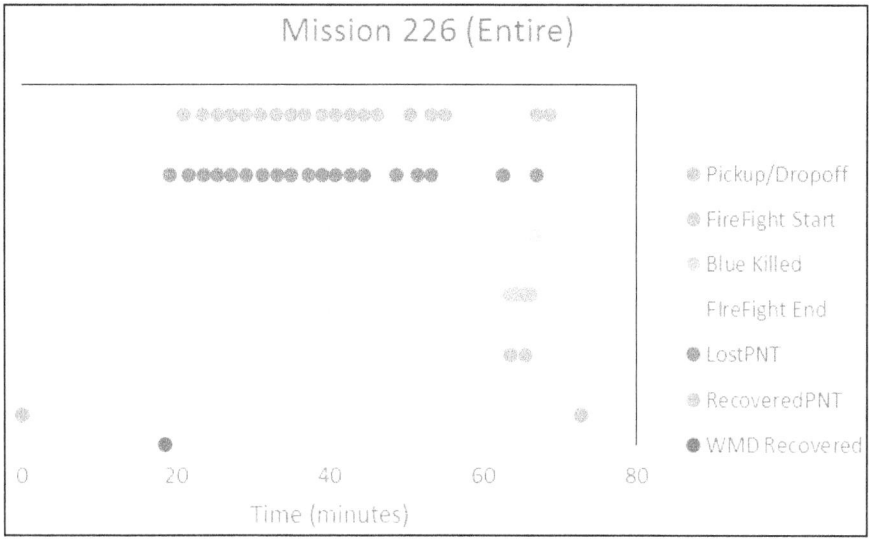

Figure 34. Individual Run #226

To check the other spectrum, we examined a mission where PNT was stressed. Figure 35 shows a low (-1) accuracy setting, a medium (0) availability setting, and a low (-1) timeliness setting. This produced a time of 872.55 minutes with 11 engagements, 44 blue members killed, and 819.65 minutes lost from GPS signal. PNT issues occurred throughout the run from start to end. Unlike the previous run, the SOF team took fire well before recovering the WMD. Figure 36 depicts the beginning of the mission. It shows that PNT issues began about 15 minutes into the mission with the first engagement occurring a little after 20 minutes from drop-off.

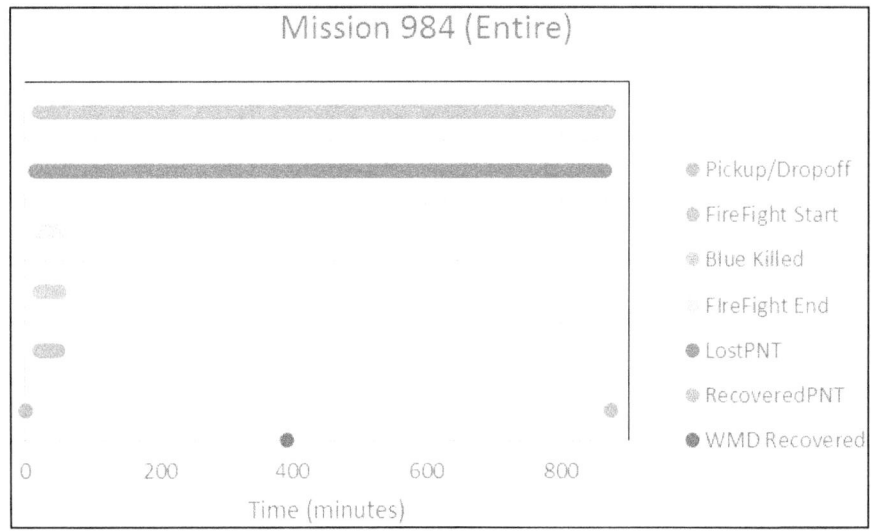

Figure 35. Individual Run #984 (Entire)

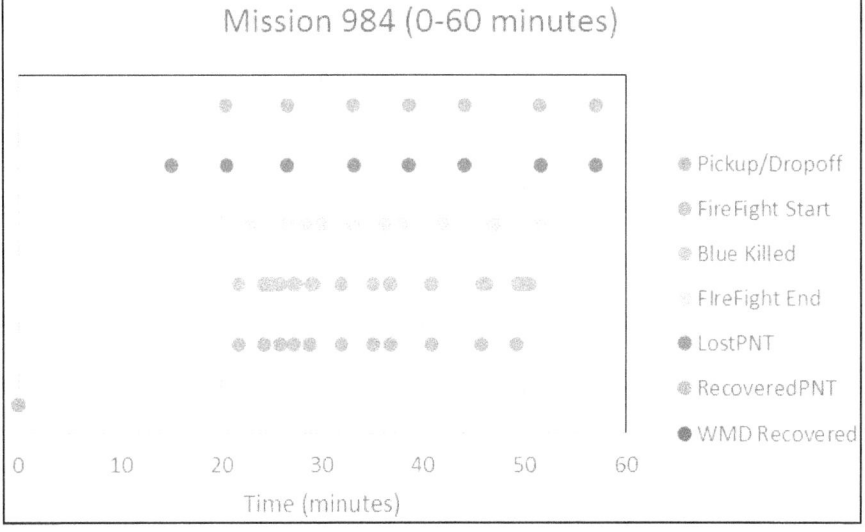

Figure 36. Individual Run #984 (0-60 minutes)

4.10 Overall Findings

This research employed a 5^3 full factorial model with eight replications for a 1000

run data set. Eighty of these runs caused erratic results in the simulation and were

removed, however all 125 data points had at least three replications. The results clearly show PNT has a significant effect on mission statistics. Of the MOEs studied, blue members killed and number of engagements did not produce a significant predictive model, which means that the three MOPs used for this research did not significantly affect these two effective measurements.

The MOE, time to complete the mission, produced a significant predictive model. This model includes higher order terms due to the exponential spike in the response as PNT began to degrade. This model contains all three MOPs. Spoofing, represented by the performance measurement, timeliness, maintains the highest individual coefficient in the model. A higher coefficient means that a given value influences the model more than a smaller coefficient. When combining the jamming effects and looking at the percentage added to the model, jamming has a slight 53% to 47% edge over spoofing. Jamming has a higher effect on the timing measurement.

The other significant model produced by the model, number of minutes GPS signal lost, also included higher order terms due to the exponential spike in the actual data. This model also contains all three MOPs. As with the timing measurement, timeliness again had the highest single effect on the model. When combining the jamming effects, jamming for a second time has the slight edge of 54% to 46% over spoofing. Jamming once more has the higher effect on the total GPS signal minutes lost.

V. Conclusions

5.1 Overview

This thesis uses a model presented at MORSS by SMC/XR to study the effects of PNT in an urban canyon environment. The overall mission involves a blue SOF team recovering a WMD from a red military force that includes local police and civilians with the help of GPS units. Three MOPs are studied, availability and accuracy representing jamming, and timeliness representing spoofing. The MOEs to understand the effects of PNT are time to complete the mission, blue forces killed, number of engagements, and the number of minutes lost by the GPS signal.

This chapter summarizes the results of the analysis conducted using simulation, DOE, and RSM statistical tools. In addition, recommendations for improvement and suggestions for future research are included.

5.2 Analysis Conclusions

The analysis clearly shows that PNT can directly and significantly influence a mission. Of the response variables monitored, the SOF team members killed and numbers of engagements responses were not statistically significant for our selected PNT MOEs. The prediction models included higher order terms due to the exponential spike in the responses as PNT began to degrade.

With the timing of the mission, all input variables were included in the model with timeliness having the biggest coefficient in the model. This held true looking at the individual variable averages across their levels as timeliness had the biggest and most definite spike, although combined jamming, both availability and accuracy, had the greater effect. The SOF team had the lowest mission time when all factors were at their highest level. Confirmatory runs validated the prediction interval.

GPS signal minutes lost is related to all factors. The prediction model includes all three inputs at varying degrees. The coefficient for timeliness once again had the highest coefficient in the model with the combined jamming having the greater effect. The GPS receiver lost the fewest minutes when all factors were at their highest levels. Confirmatory runs validated the prediction interval.

Within the two prediction models, timeliness is the largest coefficient for all models. Timeliness, representing the spoofing capability, has the largest single influence for these two responses in this design. When combining all terms in the model, jamming, which includes both availability and accuracy terms, has the larger effect slightly over spoofing for both models.

An effective predictive model was established for two of the MOEs, time and lost number of GPS signal minutes. The models produced response surfaces that generated a better understanding of the variables and their relationship to the response. For the time response, a stationary ridge was produced for the factors accuracy and availability. This tells us that no matter the level of availability, if accuracy is between 30 to 70 meters, there is little effect on the maximum time produced. The other variations, timeliness vs accuracy, and timeliness vs availability produce a rising ridge where essentially there is a slope, the higher the level of the factor, the lower the time.

For the GPS signal minutes lost, a saddle point is produced for accuracy and availability. The saddle pint is very shallow, really more of a stationary ridge. For the accuracy factor between 30 to 70 meters produces a high minute lost no matter the availability. The other variations, timeliness vs accuracy, and timeliness vs availability produce a rising ridge where essentially there is a slope, the higher the level of the factor, the lower the minutes lost.

The individual runs provided additional insight from selected replications and cemented the predictive models established. Number of engagements and blue killed did not seem to factor. With both responses, time and minutes lost, the higher the levels of all factors, the better these two responses were.

5.3 Recommendations for Improvements

As with any type of research and analysis, more data is paramount to better information. This research employed an unbiased 5^3 model with eight replications at most design points. With a larger model and more replications, a more detailed view of the model space could be achieved. This may yield a better understanding of the number of blue members killed as well as the number of engagements. In addition, a better model could be established for the timing and GPS minutes lost measurements. Continuing with the data improvements, the eighty runs lost could be ran again to better understand the design space.

60

5.4 Future Work

Many potential future routes are available to continue this research. The police and civilians in the scenario are only used as eyes for the red military. A DOE could be employed to see what effects a combat engaged police force would have in the scenario.

Another approach could be looking at different MOEs to model. Additional MOEs could be if the SOF team is detected prior to exiting the building housing the WMD, was there a change to the evacuation plan, what is the chosen evacuation plan, and if the mission was a success or failure.

References

[1] GPS.gov, "Space Segment," 02 August 2014. [Online]. Available: http://www.gps.gov/systems/gps/space/. [Accessed 06 October 2014].

[2] Navipedia, "GPS Ground Segment," 18 September 2014. [Online]. Available: http://www.navipedia.net/index.php/GPS_Ground_Segment. [Accessed 06 October 2014].

[3] C. E. Hoefener and B. Van Wechel, "P-code versus C/A-code GPS for range tracking applications," *Navigation,* vol. 38, pp. 289-293, 1991.

[4] T.-H. Kim, C. S. Sin and S. Lee, "Analysis of effect of spoofing signal in GPS receiver," in *Control, Automation and Systems (ICCAS), 2012 12th International Conference*, Jeju Island, South Korea, 2012.

[5] B. M. Ledvina, W. J. Bencze, B. Galusha and I. Miller, "An in-line anti-spoofing device for legacy civil GPS receivers," in *Proceedings of the 2010 International Technical Meeting of the Institute of Navigation*, Portland, Oregon, 2010.

[6] A. Cavaleri, B. Motella, M. Pini and M. Fantino, "Detection of spoofed GPS signals at code and carrier tracking level," in *Satellite Navigation Technologies and European Workshop on GNSS Signals and Signal Processing (NAVITEC), 2010 5th ESA Workshop*, Noordwijk, Netherlands, 2010.

[7] T. E. Humphreys, B. A. Ledvina, M. L. Psiaki, B. W. O'Hanlon and P. M. Kitner, Jr., "Assessing the Spoofing Threat," 1 January 2009. [Online]. Available: http://gpsworld.com/defensesecurity-surveillanceassessing-spoofing-threat-3171/. [Accessed 12 November 2014].

[8] J. S. Warner and R. G. Johnston, "GPS spoofing countermeasures," *Homeland Security Journal,* vol. 25, no. 2, pp. 19-27, 2003.

[9] N. O. Tippenhauer, C. Popper, K. B. Rasmussen and S. Capkun, "On the requirements for successful GPS spoofing attacks," in *Proceedings of the 18th ACM Conference on Computer and Communications Security*, Chicago, 2011.

[10] A. Pinker and C. Smith, "Vulnerability of the GPS Signal to Jamming," *GPS Solutions,* vol. 3, no. 2, pp. 19-27, 1999.

[11] A. M. Law, Simulation Modeling and Analysis, 4th ed., New York: McGraw-Hill Science/Engineering/Math, 2006, pp. 619-658.

[12] D. C. Montgomery, Design and Analysis of Experiments, 8th ed., Hoboken, New Jersey: Wiley, 2012.

[13] R. H. Myers, D. C. Montgomery and C. M. Anderson-Cook, Response Surface Methodology: Process and Product Optimization Using Designed Experiments, 3rd ed., Hoboken, New Jersey: Wiley, 2009.

[14] "TeamSEAS," [Online]. Available: https://www.teamseas.com/. [Accessed 28 November 2014].

[15] J. B. Honabarger, "Modeling Network Centric Warfare (NCW) with the System Effectiveness Analysis Simulation (SEAS)," Masters Thesis, AFIT, 2006.

[16] B. Dainty, "Position, Navigation & Timing (PNT) Study," in *MORSS 2009*, Fort Leavenworth, Kansas, 2009.

[17] Joint Staff, "Joint Capabilities Integration Development System," DoD, 19 January 2012. [Online]. Available: https://dap.dau.mil/policy/Documents/2015/CJCSI_3170_01I.pdf.

[18] General, Commanding, "Marine Corps Warfighting Laboratory (MCWL)," DoD, 2015. [Online]. Available: http://www.mcwl.marines.mil/.

[19] Joint Staff, "Universal Joint Task List," 2015. [Online]. Available: http://www.dtic.mil/doctrine/training/ujtl_tasks.pdf.

[20] E. Bland, "GPS 'spoofing'could threaten national security," Discovery News, 02 October 2008. [Online]. Available: http://www.nbcnews.com/id/26992456/ns/technology_and_science-science/t/gps-spoofing-could-threaten-national-security/#.VPB6LPnF9KI.

[Accessed 05 December 2014].

[21] D. Shepard, "Characterization of receiver response to spoofing attacks," University of Texas at Austin, Austin, Texas, 2011.